THE VOICE OF SEEING

A Toltec Father Way of Knowledge

THE VOICE OF SEEING

A Toltec Father Way of Knowledge

by Ayakel Quetzalcoatl

The Voice of Seeing
A Toltec Father Way of Knowledge
by Ayakel Quetzalcoatl

Copyright © 2020 by the Institute of Perception
All rights reserved.
No part of this book may be used or reproduced in any manner whatsoever without written permission, except in the case of brief quotations embodied in critical articles and reviews.

Cover design and Illustrations by Ayakel Quetzalcoatl
Editing by Twila Chigwa

Produced by the Institute of Perception

"The first act of a teacher is to introduce the idea that the world we think we see is only a view, a description of the world. Every effort of a teacher is geared to prove this point to his apprentice. But accepting it seems to be one of the hardest things one can do; we are complacently caught in our particular view of the world, which compels us to feel and act as if we knew everything about the world. A teacher, from the very first act he performs, aims at stopping that view."

Carlos Castaneda

CONTENTS

Prologue: FIRST STEP	19
Introduction: ANOTHER WAY OF SEEING	21

I

THE PLACE Δ WONDERLAND	33
Not-Doings	44
Not-Doing: Finding a Personal Power Spot	47
Not-Doing: Finding the Larger Power Spot	50
Not-Doing: Seeing Connections Between People	52
Not-Doing: Stilling the Eyes in Nature	54
Not-Doing: Stilling the Eyes in the City	56
Not-Doing: Seeing Love	57

II

THE PATH Δ YELLOW BRICK ROAD	67
Not-Doing: Following Colors	77
Not-Doing: Shadow Gazing	82

III

THE WISH Δ UPON A STAR	89
Not-Doing: Setting Intention	98
Not-Doing: Tricking Reason	100
Not-Doing: Seeing Your Personal Fear	103

IV

THE POWERS Δ SUPER HEROES	111
Not-Doing: Hearing As Seeing	124
Not-Doing: Crossed Eye Seeing	126
Not-Doing: The Seeing Catcher	130

V

THE GIFTS Δ WISHING STONE	137
Not-Doing: Finding a Power Object	148
Not-Doing: Seeing the Luminous Grid at Night	151
Not-Doing: Understanding the Eagle	154
Not-Doing: Activating "Will"	155

VI

THE ALLIES Δ IMAGINARY FRIENDS	161
Not-Doing: Calling an Ally with Closed Eyes	173
Not-Doing: Calling an Ally with Open Eyes	177
Not-Doing: Seeing The Red Grid during the Day with Closed Eyes	179
Not-Doing: Seeing The Red Grid, Honeycomb and Sparkles during the Day with Open Eyes	180
Not Doing: Flattening Steep Ground	182

VII

THE DIRECTIONS Δ PERSONAL WINGS	187
Not-Doing: *Seeing Your Direction*	194

VIII

THE FAMILY Δ LAST WISH	203
Not-Doing: *Seeing the Colored Dust*	209
Not-Doing: *Seeing the Colored Dust in a Person*	211

IX

THE NEXT WORLD Δ PAST, PRESENT AND FUTURE	219
Not-Doing: *Seeing Everything*	226

X

Δ THE RECAPITULATION Δ	233

Dedicated to my Three Children:
Boe, Graham, and Luke

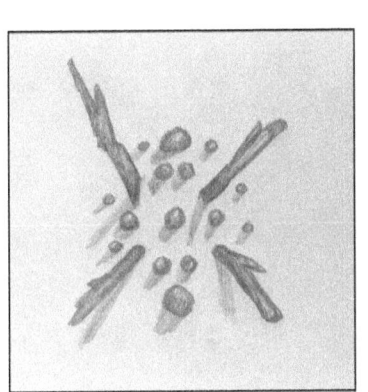

Prologue

FIRST STEP

We all remember those times when life allowed for magic. Perhaps you were around five years old when something happened that ignited your imagination. Or maybe you were in your twenties when you stumbled upon a special place in nature where you revived your freedom. Maybe in your thirties you found a special group of friends who opened the door of your creativity for the first time. Perhaps as an elder you watch an eagle ascending through the clouds, and at that moment your perception is freed from the fear of death.

Moments that allow the Unknown to dominate over the Known are magical. When the coincidences are too unusual but masterfully appropriate for the moment, so they set your world-view into a new frame of reference, then you have experienced magic. Magic is when you are no longer constrained by another's arrangement of the Play, when you are momentarily freed to write the script of your life, to sculpt your vision into reality, to believe in your own *seeing*. The particular feeling of direct knowing is *seeing*. When the experience is so complete and unfiltered that your routine world stops, then you have taken the first step off the familiar trail, and it is at that moment that you will begin to hear the *Voice of Seeing*.

This book is about remembering the time when you invented a world, when you practiced for a brief moment in a place of wonder. That moment may have occurred in town or in the mountains, but it became a place out of time and reason; maybe it was in a magnificent redwood grove, or maybe that moment was in a playground where you built a tree-fort with your friends, or perhaps that magical moment happened on an unknown street that suddenly reached a dead-end overlooking the entire city of lights.

Every parent takes the opportunity to enter into the dream of their child's imagined world, at least a few times in their growing years. This book is a return to the innocence of showing your children how you first viewed and interacted with the wonder of the world. This book is about allowing your children to guide you into their imagined worlds. The city and wild nature are available as a mysterious stage set, where time can stand still for a brief moment in our busy schedules and we do something together… something so grand that it is the day you both will remember till the end of your lives. Maybe it will be the day that you entered with your young child into a Play world that is so special that it will be one of the places you return to for happiness whenever you are lonely, sad, or in your dreaming.

Introduction

ANOTHER WAY OF SEEING

The Play enters the realm of practical magic when your intention is to *let imagination dominate reason*. You begin by voicing your magical intention, and then hold on to that intention against any obstacles that attempt to remove the veil of mystery. You transform the mundane. Only when this Play truly enters into the consensus of reality will it become *seeing*. *Seeing* is the intent to activate real magic in the physical world, whether in the streets of the city or the paths of nature. Magic is moving your perception point away from its habitual location into another syntax of explanation. The lake is no longer just a "lake", it is for example, the kingdom of the catfish people who are hiding the wand of immortality under the water.

This book is for every adult, and for every adult with young children, and for those who need a *remembering* just before they leave on their definitive journey at death. This book is about remembering those times in our life when the passageways of perception opened a little wider; palm trees were no longer only trees because they had become the tall grass that had arrived as coconut seeds, embedded in a meteor, riding on the ocean from another planet. In that creative moment, we have moved from *looking* at the trees, into *seeing* the trees.

Seeing allows for new information to flood the senses, to saturate the entire body with creativity. We are in the process of creating worlds. We enter into the space with the intention of making this moment into a magical map. The people on the street are no longer classified in the syntax of your old world, but become keepers of omens, purveyors of unusual foods from strange galaxies, and appropriate guides to usher in your new world context. In this process of interacting without fixed definitions from the routine world, you become a perceptual magician able to clearly *see* outside accepted glosses of interpretation. A perceptual magician is someone who *sees*.

Seeing holds the intent of using the whole body to detect the hidden. Although the eyes are the primary sense used in *seeing*, it is the entire body that responds to the opening of the larger field of perception. *Seeing* is the creative activation of the entire body to move into its totality. We *see* more; we *hear* more; we *feel* more. When we *see* there is a moment when the chaos of debating voices in our head suddenly focus with pinpoint accuracy, and whisper truth. At that moment you hear the one voice that has always been clamoring for dominance. It is the true voice. That is the moment when you realize that you have been always hearing the Voice of *Seeing*.

The Voice of *Seeing* can only be accessed when you have reached inner silence. When the Voice of *Seeing* arrives, after years of disciplined *seeing* of yourself, you will have no doubt that its words are simple validations of your purest essence. It is a soft voice. It suggests, it never demands. You will recognize the Voice of *Seeing* by its poetic simplicity. It speaks to your soul, and your soul acknowledges its profound understanding of any situation. The Voice of *Seeing* is a direct link with the intent of *seeing*.

In the process of participating with a magically charged world, we eventually acquire practical tools of perception that last a lifetime in the routine world. We pass these tools on to our own children, and they in turn share the lineage of magicians with the next generation. Creativity is the beginning of any world. The skills learned in the realm of pure creativity are the tools that are used in the pantheon of adult professions: artist, doctor, or the engineer designer of a spaceship in the future.

One such tool is the concept of the *not-doing*, which is at the core of the magical tasks at the end of each chapter in this book. These tasks do not emphasize reason, but activate the idea of using your senses to disrupt the *doings* that dominate your regular activities. *Not-doing* occurs when you dedicate a particular time in a particular place to do something that brings forth a result connected to the mysterious side of being human. *Not-doing* is a concept originating from the Toltec Shamans.

In this book, I will share the first gifts from the perceptual magicians of antiquity, called the Toltec. Long before the invention of toys with wheels, or video games, the Toltec shamans used their creative time to collect information about *seeing*. They ran through the fields discovering the Kingdom of the Birds, or swam in the lakes to visit with the Kingdom of the Fish, or walked their villages to pay tribute to the man who discovered how to shape obsidian. In this book we return to the way the first humans explored their world. It contains methods to revive the love of place within you and whomever you share these special moments with: your family, a lover, a good friend, an elder, and of course your children.

Imagine that you are the father of a young daughter, who is being guided in all her introductions to the world by you. You are responsible for her day-to-day care, eating and learning to interact with other people. But in this process you also want to impart a system of ancient knowledge that will help her become not only a functioning adult, but also a creative being. You do this naturally, since you are a perceptual magician. It is your natural ability to love your daughter and give her the truths of perception; to give her the widest possible avenue for proper decision-making.

The process of *seeing* is a creative pathway toward awareness. Guiding your children to awareness using the tools of a perceptual magician is the paramount activity for a Toltec father. When you accept your role as a Toltec guide, you become your child's benefactor bestowing the gifts of perceptual *seeing*. That abstract part of your own nature, called the "hidden-self" by the Toltec, once uncovered, completes the totality of a real human. When you have merged the lines between your two halves, the *looking* half and the *seeing* half, you have removed the mask of hidden knowledge, and can reveal in the classroom of the world, both natural and manmade, the totality of being human to your children.

You are entering the world of perceptual magic as originally chronicled and passed on by the ancient Toltec, the artist/mystics of MesoAmerica. Many of the concepts they collected during the earliest years of shamanic sciences and the terms derived from their explorations are used in this text. The Toltec were the first to define *seeing* as not looking. The understanding of Power Spots, the Energy Body, and the Assemblage Point are part and parcel of their intent.

I am continuing the Toltec tradition of imparting perceptual knowledge in an art form, in this case, writing and illustrations. Although this knowledge is from the time of antiquity, its value is not diminished by our contemporary frame of reference. On the contrary, as we move further and further from the natural world, it becomes urgent that we return to the purity of nature on occasion to revive our human *seeing*. Stopping your customary perception of the world is the first act of *seeing*.

The perceptual magicians of antiquity sought out apprentices through omens that occurred in their day-to-day life. When they received an omen that signified a future apprentice, they jumped at the opportunity to guide them into *seeing*. As is the case with many shamanic traditions, the offspring of a perceptual magician is the first to be watched for omens that may predict *seeing* abilities. Fathers were closely watching the actions and decisions of their daughters and sons. They realized that the possibility of bloodlines of magicians were the fastest path to apprenticeship.

The Toltec used the natural world to define themselves as human. The five predominate senses and their multitude of nuances, connected primarily with the sense of sight, is the basis for our understanding the concepts of Toltec magical viewing.

Imagine that you are the teacher presenting the knowledge of *seeing*. You will be the "father" character, in these tales of power, who has taken the wand of knowledge and bestows the gift of *seeing* to his three children apprentices. The *"you"* in this book is the *"you"* of every person who has undertaken the challenge of *seeing*. You have walked this path from birth, and I, as the Voice of *Seeing* in this particular text, am only reminding you, that you have always known these magical tasks… and you only forgot your own Voice of *Seeing*.

Walk with me. I want to share with you, what I have remembered about creative Play in a safe environment, but in this process, I will reveal real practical methods to *see* beyond the Known. This book is written from the perspective of a father who is walking through worlds with his young children, and who is maintaining his own child-like wonder. He is listening to his inner counsel... the Voice of *Seeing*.

The Voice of *Seeing* is a direct link to the intention of the ancient Toltec magicians. It has been set into play long ago by those perceptual magicians of antiquity. It was and still is, their original intent guiding these lessons. The Voice of *Seeing* includes the lessons the Toltec accumulated from their first *seeing* of this beautiful Earth... when the wonders of the water and the marvels of the air were glimpsed for the very first time. You are walking into the past, and then into the future at the same time. This time-travel is at the essence of moving your perception out of a fixed position. It is a simple and yet a complex text that bridges the Known and the Unknown... the physical and the abstract nature of perception.

Everyone makes *wishes*, but Toltec magicians would *see* the answers in omens that manifest in the world, and hear validation from the Voice of *Seeing*. Omens are signs or symbols or objects in the outside environment that pertain to your voiced *wish*. The Voice of *Seeing* is the eventual clear and trustworthy voice inside your head. Its sentences are poetry, and its tone is forgiving. It will resonate with everything that is important to learn.

The *wish* is the intention voiced at the very beginning of your walk that needs to be answered through *seeing*. The *wish* must be delivered out loud in one clear sentence, for the answer to be delivered by a surprising cast of characters waiting along the road. When we *wish* upon a star, we are intentionally projecting our sincerity into a tangible shiny object; we are taking the abstract and attaching it to material reality. Our *wishes* are bridges between worlds.

The answer to your *wish* may come from an omen, like a stone in the shape of a heart laying on your trail. The Voice of *Seeing* will tell you what this stone means for you. The answer to your *wish* may come from an omen that is a rock ledge supporting you. The Voice of *Seeing* will tell you that the Earth will support you forever. The Voice of *Seeing* is your internal compass and a validator that you are walking the proper path in the proper direction and *seeing* your answers. When the voices of reason and fear are momentarily silenced, then the Voice of *Seeing* will clearly answer all the questions about what you are experiencing. Trust your guide.

We are first and foremost perceivers of energy; we are magical beings. You are a perceptual magician walking in a mysterious world. I will momentarily be your guide, but when your *seeing* comes alive, I will gladly pass this found branch over to your daughter, who is waiting by your side to guide both of you, because she understands instinctively that it is no longer just a small stick, but the *wand* of immortality.

When I began this book, I was in a park in Israel. It was the very first time I had visited this urban park between railroad tracks and a freeway. I felt at home. It was a very hot day and the song of the flowing stream pulled me over to splash water on my face. Then I saw a nearby lake formed from the stream. When I got there a family of large catfish came to greet me, while a bird dove under the water for its dinner. The trees waved hello, the small thorns stuck to my skin as a blessing, and the moon was rising over the nearby ruins of a Roman fort.

It was no longer just a park. I was in need of a special space, a Power Spot, a place of the imagination, where all the players aligned with the intention of being their totality, both real and invented, and the only rule for this game of worlds was creative freedom. Finding a natural Power Spot is my first priority when I plan to stay in one location for an extended time. It will be your first act in our walk together. A Power Spot is an essential ingredient for successful *seeing*.

My grown daughter asked me once what was the one memory I would take with me to infinity after I died. I answered that it would be one of the magical times we had together when she was just beginning to view the world and I was carrying her in a backpack on my back. We were walking through the forest and she was stretching her legs by bouncing in the backpack to see over the top of my head. As we progressed she was *seeing* all the trees for the very first time, and was speaking to them with her first word…"Hi".

It was not only a memory of a fun time together, or even a nostalgic feeling of adult and child long gone; the memory still invoked in us a feeling of magic time. It held a perceptual intent that still had the power to make our bodies long for magical moments. It was magic because my daughter was *seeing* for the very first time in such a pure form, that the trees were living-beings who could be acknowledged with a "hello". It was magic because memories like this are the ones we choose to keep with us for eternity.

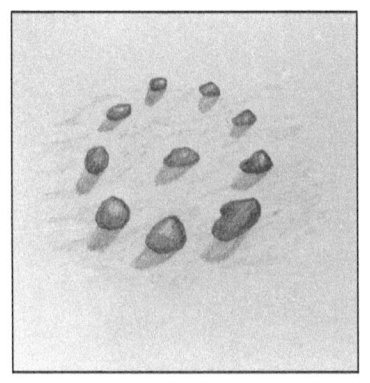

I
The Place: Wonderland

When the Toltec apprentice was taken for a walk that involved the development of their awareness, they underwent a profound change in their perception of the universe. They were instructed to stop their world; the routines of their day-to-day life were put on hold, in order to make room for another way of viewing the world, which they labeled *seeing*.

The absolute necessity for developing their perception required, first and foremost, finding a Power Spot in nature. The Toltec had found many in the mountains, and took their apprentices into their protective radius. In places of power, the chance of finding and *seeing* magic was enhanced. Just like with any discipline, the gift from the teacher was that they knew the qualities to look for within the landscape, for they had also been taken to a Power Spot as a young apprentice. It was a place of unrestricted freedom that they were introducing to the apprentice. It would be the classroom for the studies in learning how to *see*.

Wild nature held all the lessons of the water and of the wind. In this classroom there was unlimited variety in every tree and textured stone. When the Toltec apprentice *gazed* into a moving body of water, a mysterious transference of energy occurred in their *seeing*. When they observed a waterfall, the movement of the falling water was retained inside their eyes, and could be transferred to the surrounding landscape when they shifted their eyes to the nearby mountains. The mountains began to move like the water. The ability for moving water to make the solid world into a river of flowing energy was the key to successful *seeing* for the young apprentice. This not-doing lesson showed the young apprentice that the world was forever changing, and that to be attached to one perspective was the real illusion.

The great illusion was the human illusion that everyone should *see* the same world and interpret visual information in the exact same way. The apprentice was instructed to clearly view the world outside agreements made before they were born; the traditional methods that corralled all humans into the same worldview. It was the teacher's intention to move the apprentice into a creative view that removed them momentarily from the accepted norm, and in this process opened the door to untainted awareness.

The teacher would position the apprentice to view village life from a distance, which allowed *seeing* the similarity in everyone moving through the daily activities. Although someone might be preparing food and another person collecting firewood, everyone in the community was moving within the same energetic framework, which the Toltec named the "assemblage point" or the "mold of man". They understood that humans were bound together for survival as part of their mold. They all intended to work and play together for the general good of their tribe. But in addition to the temporal duties of daily life, the Toltec were fascinated with the great abstract desire for humans to find an illusive feeling and deep connection with another, that they named "love".

The Toltec understood the connections between parents and their children, between loving couples, between the bees and the flowers, and even the clouds and the rain that loved the Earth into fertility... but the illusive substance of love was what they wanted to *see*. They sent out teams of apprentices to record every interaction that showed the qualities of love. The apprentices returned with many examples of love to the council of wise men. They could *see* love exhibited between a man and a woman, and for their children, they could *see* love in the eyes of a man who had just finished a finely shaped arrowhead, and they could *see* kindness showed to animals. One apprentice had climbed to the highest mountain and was sure that the ultimate expression of love was a rare blue butterfly he had observed leaving its cocoon and *seeing* a flower for the first time.

There were as many examples of love as there were apprentices. Finally the teachers went deep into their dreaming in the search for love. They sent their Energy Bodies, or their ethereal bodies used in dreaming, to the far corners of the universe, landing on distant planets and exploring the flora and fauna of other worlds. After all the research had been collected, the counsel held a meeting on the steps of the Quetzalcoatl pyramid in Teotihuacan. They announced that they would reveal the end result of their combined search for love. The youngest teacher walked to the center of those assembled and said:

"We have finally found love. It is a rare and beautiful substance that shines from the eyes of everyone who has reached the source. In all the plants, in all the people of this world and all the other worlds in the sky, there is one substance that unites everything. Love is awareness."

That day the council not only revealed the final distilled essence of love, but they also could *see* in this young teacher the perfect representative of love. They named this person a Nagual, a person of knowledge, the ultimate guide to *seeing* and someone who could truly love in the biggest sense of bestowing awareness to another.

....

Now you will be entering into the gaming field. You are prepared to take on the character of the "father", who is guiding his daughter apprentice into *seeing*. This is your first walk together; each subsequent chapter will be another journey further into the mysteries of a world just over the horizon. You stand at the threshold of magic, holding your young daughter's hand. And now… you take your first step forward.

You are walking with your young daughter in a large city. There are skyscrapers and cars and noise all around. You wander over to a small park in the midst of all the activity and it is here that you find a comfortable home base. You wonder if others feel more secure inside one of the tall buildings, perhaps at a crowded restaurant or even in a car listening to their favorite music. But you have made an intention to interact deeply with the outside environment; you just needed to locate a small place of natural solace in the middle of your gaming field.

The city is an unlimited game board. All the people are players in a complex script with mysterious overlaps. From where you are sitting, you have a view of an open field bordered by tall pine trees. The gently sloping lawn gives way to the flat top where you and your daughter feel protected within a wall of sculpted hedges on three sides. It is as if the landscape artist intended for people to sit within this sheltered island of tranquility, in order to meditate on the panorama of the greater park. You are here to make a *wish,* and you say to your daughter:

"I *wish* that we *see* love."

You sit together in silence and wait. You are waiting for an omen; something occurring in the outside world, a sign or an event, that brings on the feeling of *love*. You watch all the people go by, and attempt to realize through your *seeing* if they are connected in anyway with love.

When we intend to find love we usually connect that word with other people. We *wish* for a loving partner, a loving best friend, or someone who loves life as much as we love life. But today you are intending to unveil its abstract nature, that part which is involved with magic. Love is the essence of life, and yet we only *see* the physical reactions after feeling love: a smile, a kiss, or a material present. As a Toltec father guide, you are *wishing* to *see* love as a magical moment of energy flowing through the universe and uniting two separate beings. You say to your daughter:

"I *see* people who share."

You are referring to the street vendor who is holding a handful of balloons, and to the group of laughing children running through the water sprinklers, and to the passing adults who are holding hands. You get up and go toward the vendor to purchase a red balloon. You bring the red balloon over to your daughter. Then she passes the balloon over to one of the other children in the park, who accidentally lets the balloon fly into the sky. The red balloon is carried by the wind over to a teenage couple holding hands; they reach up at the same time and catch the red balloon. You say to your daughter:

"Sharing positive energy is love."

You walk out of the park into the city street. Every person is actively going somewhere. You both stand still in the midst of the stream of people. Everyone is moving past you on all sides. You feel almost invisible, like two protruding rocks in a rushing river. You instruct your daughter to *still* her focus at the center of the passing people, going toward or away in all directions. You both hold your eyes *still* in one area of the crowd, focusing at feet level for a long time, in one spot, never being tempted to follow any one person. In the process, your own internal dialogue is silenced, and replaced with a separate voice. This voice is a whisper of poetic truth and clarity. The Voice of *Seeing* says:

"Seeing is the mastery of silent knowledge. My voice comes from the center of the universe and flows through all beings that have wished to hear awareness speak to them. You are now seeing the unity of life and hearing internal validation at the same instant."

Your daughter tugs at your hand and tells you that everyone is part of a big river moving all around you. She understands that being still and focusing on one spot has stopped her old perception of people and allowed her to imagine them as a gigantic river. She is *seeing* the tide of humanity. They all seem to be on a conveyor belt, a large river of unified intention. Everyone is connected to the other with a pattern of movement that matches in cadence and rhythm. They all move with a style and general speed that unifies their randomness into a cohesive whole. The Voice of *Seeing* continues:

"You are seeing instead of looking. The world of humans has a pattern that all humans adapt with their doings. You are now seeing a moving example of the human assemblage point. The human assemblage point is the abstract point of energetic agreement to be a human."

You break your focus and go back to your park home base. It feels good to be back in your protected space. The wall of leafy hedges forms a safe wall around your personal spot. Your daughter picks a few daisies and arranges them in a small ring around your sitting area. She then adds a few small stones that equally divide the radius into four quadrants. You sit in silence within your circle for some time observing the outside world. A group of small birds dot the lawn; they all seem to be united in their search for seeds or worms. An airplane crosses the sky carrying a unified group of travelers to another land. You imagine the sky uniting the entire Earth in a circle of blue. And then, at the edge of the park you focus on a real river of flowing water that unites every living being with the essence of life.

As you stand by its shore, the moving waters invite further explorations with your *seeing*. The water glides swiftly through the landscape and the sun's presence unites the ripples with sparkles of bright light. The river invites a perceptual shift of attention toward silent knowledge. The Voice of *Seeing* says:

"You are a perceiver first and foremost. Seeing unites with nature to form pure experiences outside of any internal debate. Being able to reach the assemblage point of water is to move beyond human reason and ride on a new current of perception."

You ask your daughter once again to *still* her vision in one small section at the center of the fast moving water. She is tempted to follow the water's direction, but she resists the temptation and stays focused at one point in the moving stream. Then you ask your daughter to suddenly raise her eyes to a point at the middle of one of the skyscrapers above the park trees. You instruct her again to *still* her eyes on one section of the building wall and wait.

Magically her eyes retain the movement of the water. The solid exterior of the skyscraper now mysteriously moves like the river; transformed by the fluid movement of the water. Its walls move with an overlap of ripples and watery fluctuations. Your daughter has transferred the action of the river to the building wall with her *seeing*.

The ancient Toltec could *see* that water, the essential ingredient for life, was able to flow into the structure of all living things on the planet. They also could *see* that water was more than a life giving substance; it was a tool for *seeing* the invisible transparent nature of energy on the planet. Water represented the moving force of the spirit. Your daughter makes a wondrous statement:

"Water loves everything."

You reflect on the nature of your own being that has changed, now that you are walking with your young daughter. For many years you were all alone collecting perceptual information, and now you have someone with whom to share your discoveries. You are her teacher, as well as her father. You are teaching her to *see* the makeup of the universe as flowing energy in addition to her learning the proper etiquette of everyday life. As a Toltec father you have the responsibility to continue passing the gifts of perception to the next generation. Then you hear the Voice of *Seeing*:

"Perceptual magicians are a bridge between the Known and the Unknown. They are perceivers and collectors of magical events in both worlds. The makeup of the universe is a flowing stream of energy. Seeing is an energetic force that motivates action in a positive direction... just like love."

As you walk home, you wonder if the day has been as magical for your daughter as it has been for yourself. You concentrate on the sidewalk stretching out before you, a path that can be transformed into an aerial landscape as viewed by a flying eagle. You *see* yourself and your daughter as two eagles flying over the sidewalk far below. You *see* the duplicated paving stones as agricultural fields of growing crops separated by lines that are miniature fences. You take your daughter's hand and fly faster and faster down the street into the next world. The Voice of *Seeing* whispers:

"Love is sharing Awareness."

NOT- DOINGS

The not-doings are instructions at the end of every chapter to help you personally experience the magical events described in the previous story. The not-doings can be performed individually or with your children. If you are guiding your children, inspire them with a creative presentation, custom tailored to their life style. You always match and relate to your children from their perspective. You are moving your own perception to meet with their budding perception.

The not-doings provided, should remain fluid and adaptable to your particular personality, and your children's life style. In addition, the not-doings should remain flexible for any situation you may find yourself, during trips to your park or city streets as well as wild nature. You are the guide leading your children into the world, and your creative method of presentation is a direct reflection of your temperament. If your children react best to a not-doing introduction that uses a recent event from their own lives, then you seize the chance to incorporate that event into the process. You are teaching them creativity through your own creativity in the moment. These are not-doings because your intent is to change the usual methods of presentation we are accustomed to in schools, or in the work environment. The Play of the not-doings is a matter of impeccable timing and perfect place. Pick your game board carefully when all the elements are aligned for success.

Example of Not-Doing:

My eldest son, barely six years old at the time, liked to play with action figures and construct a story based on the castle we had constructed with wooden blocks. During one of our Plays together he stopped the action and asked me, "Why do the bad guys always lose?" I was taken aback by his astute question. I realized that we had been in a Play that conformed to every cartoon or book or movie he had ever seen where the bad guy always lost in the end. He had guided me through his own budding perception into the frozen nature of all the Plays ever written. That day we let his villain action figure win the struggle to capture the castle. The protagonists ran for the hills to perhaps fight another day. My son had created a "not-doing".

Follow your own Voice of *Seeing*. The Voice of *Seeing* is an internal suggestion for a greater understanding of the external events, but it relies on your personal intention and creative discipline for a complete adaptation to your life and your children's life. The not-doings are the magical practices of a perceptual magician. They have been developed by the ancient Toltec, who listened to their own Voice of *Seeing* in order to arrive at these not-doings, and then continued clarifying them to perfection over generations of intended magicians.

The not-doing magical tasks are your birthright to be a complete perceiver of your beautiful world. Walking within the Play of your world, without *seeing* your full potential is partial blindness. These exercises are not difficult nor too complex, because they are the natural extension of your latent perceptual abilities gifted to you at birth. Consider yourself as a perceiver, who although momentarily blinded by years of distraction caused by the chaos of your Known world, now has the chance to have a chance at total *seeing*. Use these not-doings when the moments of aloneness or sadness attempt to overtake you. Use these not-doings to thrust you into remembering the entire range of your gift of perception. Remember that *seeing* is one of the tools you use to decipher correctly your walk upon this beautiful planet.

Perceptual magicians saturate themselves with a single not-doing for a period of time, before moving on to another task. The Toltec defined not-doing as a rigorous unfamiliar act that saturates reason to the point of submitting to a new attention. These are not-doing practices, so remember to be flexible and not develop a fixed routine of *doing* around the tasks. If you master one not-doing per week, then you remain flexible. The energetic field of a perceptual magician should not be depleted with "doing" exercises, but invigorated with the ease and creativity that results from engagement without the rules of acceptable reason and frozen routine. The normal routines of life take vast amounts of energy to complete. We are all involved with the un-acknowledged condition of constant "doings". For most, the idea of a brief vacation is also a "doing" with necessary planning and rehearsed activities. From school to work to retirement, our life is full to the brim with "doing".

Be prepared to enter an ancient system where doorways to new perceptions are opened. Once you have moved through the door of "not-doing" your realm will widen to include time without structured goals or organized fun. Freedom is another word for a "not-doing". But this freedom is also an abstract discipline that remains grounded to an activation of your totality as a creative being on the way to complete awareness.

The following is the set of not-doings related to the story in the first chapter, The Place: Wonderland. They are the moments in the story where situations were activated and you were introduced to *seeing*. You have been prepared by this first story to actively take your next step on the path of *seeing;* practicing the not-doings.

....

Not-Doing: Finding a Personal Power Spot

Go to a place were you feel removed from the *doings* of the world. It must be connected to nature, either in the wild landscape or in a special urban park that invites contemplation. If this location is to become your Personal Power Spot you will receive an energetic boost; a feeling of totality in mind and body. As you walk through the landscape, shift your vision quickly from spot to spot without focusing too long on any area. Eventually you will be able to detect an energetic relationship between yourself and a specific spot within the general larger area. This spot feels different, it may look more peaceful or more beautiful to you, but you will only know for sure if it is your personal spot by interacting with it. Walk around and feel the perimeter and determine the very center of the area. Sit on that spot for a brief moment. How do you feel? You must feel energized in this spot. *Seeing* your personal spot is feeling that you have reached the exact perfect area for your being. It is your Personal Power Spot. It is not connected to your particular mood that day. It will always invite you to visit, no matter if you are sad or happy. It accepts everything about you; it is ready to present gifts of sound and vision every time you visit.

Locate stones or branches or feathers or any other items in the area that evoke feelings of personal connection. The objects could be associated with memories from your childhood or address the current *wish* you made at the beginning of your walk that day, or simply choose beautiful power objects that remain mysterious.

Use your collection of found items to mark a small circle perimeter around your Personal Power Spot. With this circle of objects, you have created a protective line that separates the world on the outside from your personal world on the inside of the circle. In wild nature, use rocks, feathers or branches. In the city you can also collect human tokens; objects that have been lost or discarded to mark your perimeter. When you have completed your circle, stand back to observe and reflect on the beauty of your arrangement. Reflect on the feeling of surprise that another adult, or a child and their parent might have when discovering your beautiful creation.

Sit or lie down at the center of your defined circle in your Personal Power Spot, and close your eyes for a few minutes. *See* how you *feel* with your eyes closed and if your choice of location is perfect. Let the outside sounds attract your attention and silence internal dialogue. Then open your eyes. You should experience *seeing* the outside world with a fresh vision. Perhaps you will *see* things in the environment that you missed before. You should feel energized. You should have a sense of well being when you are in your circle. You gain insights, and you feel that the world has granted you an experiential gift of personal power that will last a lifetime. When you return home from your Personal Power Spot, the trail back is joyous. You feel lighter, and the troubles and hardships of the world have been vanquished momentarily. You are joyous because you have struggled to find magic and have been rewarded with awareness.

When I found my first Personal Power Spot I wasn't even looking. I was twenty years old and walking through Golden Gate Park in San Francisco. Just past the Japanese Tea Garden, I suddenly felt compelled to change my normal route home and take another smaller path. It led me to a large field with a pile of perfectly cut building stones that were to be assembled into a monastery, which was originally erected in Spain. The sacred building had never been rebuilt in this new location. These giant blocks had remained there for many years and were now firmly embedded in the natural landscape; overgrown with weeds and even small woody stalks of trees broke through the cracks between the blocks. On the other side of one wall was a sunken arena that invited my further investigation.

I sat on one of the stone blocks. Its face had been carved with ornate lines and flourishes around a central cross. I was sitting in the ruins of a holy place. The longer I stayed silent the more reverent I felt. For the rest of the day, I cleaned away the weeds and manmade debris around the amphitheater. I collected special stones and flowers to place on one of the blocks for an altar. It was in this Personal Power Spot that I began to *see*. As I *gazed* into the canopy of swaying trees there was a blue mist emanating from the tips of the branches. It appeared as a fog that eventually enveloped the entire arena. Everything became silent. The world had stopped. I entered into what the ancient Toltec called the second attention, the place of silent knowledge and the location of the Voice of *Seeing*.

Not-Doing: Finding the Larger Power Spot

When you *feel* satisfied with your Personal Power Spot, expand your territory. Take the time to walk in each of the four cardinal directions to find the outer boundaries of your Larger Power Spot. Feel when the energy creates an invisible boundary that portends an entrance into another frame of reference; a feeling that this is as far as you should walk in each of the directions. You have now mapped the boundaries of your larger playing field. In wild nature the boundaries may always be expanding and offer an unlimited landscape. In the city your game board will most likely be a smaller area, perhaps where the park stops at a busy street.

Both wild nature and the urban environment are filled with unlimited possibilities. Over the years you will return again and again to your Personal Power Spot. You will explore all the smaller nuances within the Larger Power Spot area. You will begin to *mark* areas where the Voice of *Seeing* tells you that someone else will enjoy your offerings of art made from found objects in the landscape: stones, sticks, feathers, beach shells, a lost bracelet, one earring, all made into a mandala. Your Earth art represents your particular temperament and the places where your *wishes* had been fulfilled.

When I found my Larger Power Spot in wild nature I was in my thirties. I had climbed to a high ridgeline in Northern California with my wife and my newborn daughter. We reached the top, and a vast panorama opened before us. We were at the threshold of a valley that invited exploration into mysteries that lay hidden in the towering rock walls lining the nearby horizon. As we proceeded into the space the wind suddenly blasted our threesome with a force powerful enough to pick us off the ground. My wife and daughter took safety in a cluster of bay trees while I proceeded further into the landscape. That day was magic. I was met by a herd of wandering cows that led me down into the canyons where Condor birds sat perched on the gigantic stone outcroppings and deer meandered in the grassy spaces between these islands of strangely shaped rocks. It was here that I found my cave, my Personal Power Spot in the larger area. As the years progressed I named this wind sculpted cave my "studio". Its walls were pockmarked with cupped indentations in which I used to place my drawing materials. I walked this ridge for over thirty years, and the wondrous gifts I received from this Larger Power Spot sculpted me into a perceptual magician.

Not-Doing: Seeing Connections Between People

Go to any place where there is an abundance of people passing by in a large area. This could be in a city street, or even viewing the panorama from a bench in a busy park setting. Position yourself so that you can observe the throng of activity for a period of time. *Wish* out loud that you want to *see* the connections between the people moving around you.

Pick any one person in your field of vision that catches your attention. They will be the "human template", the blueprint for *seeing* the answer to your *wish* to *see* connections. *See* the way that person moves; if they are carrying something unusual, or if their body language reveals a hidden injury or a distortion or peculiarity in their movements. *See* what they are wearing or if you sense that they feel sad or happy. If they are a couple, *see* how they are interacting with one another.

The first person you are *seeing* will eventually lead you to the next passing person; they could cross paths when walking, or have a similar colored jacket that shifts your *gaze* onto this next person. Consider people as passing the baton of similarities from one to the other. Observe everything about the next person connecting them to the first person; the attitude or perhaps the age that in some way mirrors the first person. Your system of cataloging them under age, or height, or clothing, pertains to your original *wish* of *seeing* connections between people. When you have reached your limit of *seeing*, you can put your story together. *See* how they form a continuous script, a Play where none of the actors knowingly rehearsed their script... but you as the director have *seen* them all perform perfectly together.

You will now put all the pieces of your *wish* puzzle together. Remember as much as you can about your *seeing* of all the people. The Toltec magicians of perception called the internal review of people and events involved with their *wishes,* a *recapitulation.* They developed a simple ritual of breathing-in the memories that pertain to their *wish,* and then breathing-out any unnecessary information while exhaling. The *recapitulation* will give you the answer to your beginning *wish* by activating a visual map of all your tracking information.

The greater purpose of this not-doing is for you to see that people are part of a larger framework of "doings". As a perceptual magician you are moving away from this framework and becoming aware of things that had been unavailable before. Until you have stopped and taken the time to intend a new way of *seeing,* mysteries would have remained vague memories from childhood, ordering your world into some form of partial map. Now with your new way of *seeing*, the map can be completed.

Not-Doing: Stilling the Eyes in Nature

Go to a river or a waterfall where the water is moving consistently in a single direction. The movement can be vertical as in the case of a long waterfall, or moving horizontally as is the case of a large river. The body of water must be at least as wide as the width of a city street. The width is important because it has to flood your vision with a consistent direction of movement. A small stream will not be sufficient. This not-doing requires a good amount of visual information in order to be effective.

Sit or stand at a position across from the water, so you can observe a large section of the water movement. Keep your eyes *stilled* at the center of the moving stream of water. When I use the word "stilled" I mean a relaxed *gazing* without following any movement other then the small area of extreme focus. Do not follow the action or the direction of the water. It is your *wish* to hold your vision at one point inside the moving water and continue to hold your vision on that exact spot for a few minutes.

After a few minutes shift your eyes to any solid area by the water; to the nearby bank of Earth, to the trees, to the distant mountains or city buildings. Hold your *stilled* eyes on this new area for a few minutes. If you are successful you will transfer the movement of the water in your eyes to the secondary solid location. *See* how the solid world becomes fluid. The Earth moves, or the buildings now move with the energetic movement of the water. Note: the movement within the solid environment will move in the *opposite* direction of the water.

Most people experiencing this phenomenon find it extraordinary that they have never witnessed the water move the nearby stationary grounds through their *seeing*. After they experience this *seeing* their world is forever changed. They begin to wonder how many other things they have missed and how this simple not-doing has revolutionized their perception of the world. They realize that all the time their eyes were capable of such a feat, and yet somehow their mind erased the event as irrelevant. It is a perfect not-doing for beginning the questioning of how they viewed reality from a limited perspective.

Not-Doing: Stilling the Eyes in the City

Position yourself in an area where there are as many people as possible walking in all directions. This occurs at most places of entertainment or commerce in a city: busy shopping districts, congested city streets or festivals with hundreds of attendees. Position yourself amidst the activity so that you have a wide view of the people and then *still* your eyes in the midst of their actions, about mid to lower level in the commotion. Maintain your *stilled vision* in one area. It is sometimes easier to focus on a non-moving point like the ground they are walking upon, than trying to hold the view in abstract space at the center of all the movement. Perhaps you keep your eyes on a certain stone on the ground, or a line in the pavement, or even a stable trash can in the middle of the activity.

Now keep your *stilled* eyes on this target location or object, within the landscape of moving people, for a few minutes. Your peripheral vision should have a large enough radius that you can "catch" the lower half of many moving people framed on the sides of your vision. Keep your *stilled* eyes on this one target point. Do not follow any individual's movement.

You will begin to *see* that the movement all around your peripheral vision has a unified field of simultaneous speed and rhythm. The people and their walking will *shift* into a united pattern with a similar pace and style of walking. It will appear as a slowed down film that has been choreographed. You are *seeing* that people are stylistically matched through the agreement to be humans since the time of their birth. They naturally calibrate their energy into a general unified field of relatedness. The Toltec named this phenomenon the human *assemblage point*; or what it means to be part of the same intention within the mold of man.

Although this not-doing takes intention and enough time to affect one's *seeing*, it is well worth the effort. Just as the not-doing of *stilled* eyes in the moving river showed that elemental movement in nature can be transferred, this exercise reveals details about our humanness. When we begin to *see* the similarity of human movement we are opening the door to understanding who we are as a species. We are inherently no different from the birds or other mammals with their rehearsed customs and style of interacting. The importance of this *seeing* to a perceptual magician goes beyond the sociological implications, and right to the heart of the one difference between animals and humans. We have the distinct ability to move out of the accepted assemblage point of our species with the process of *seeing*, and this is not available to any other species on the planet.

Not-Doing: Seeing Love

As their studies deepened, the Toltec apprentice began to *see* beyond the Known and into the mysteries of the Unknown. They could *see* the radiant nature of the human body outside of the physical frame. They could *see* that every person's body was surrounded with an energy field of amber light. In addition the Toltec were able to *see* a brighter spot of light on this luminous shell. This bright spot was the point where the universal emanations had coalesced into a unified intention that they named the assemblage point within the mold of man. It was the point where perception originated and was in the exact same place for all people. Our actions are learned and silently agreed upon with the constant training from the adults who taught us about their world since our birth. The assemblage point of being human was frozen in that location and only through perceptual intention could it be dislodged for a greater range of *seeing* and move to a different location in the mold of man.

Each generation of children grow within a defined frame of reference that shifts back and forth between the dualities of this planet, but it is always within the framework of the human assemblage point, within the mold of man. The Toltec taught their young apprentices methods to move their assemblage point beyond this duality and called that new location of the assemblage point *seeing*.

The continual search for love has always been part of the common human assemblage point. The arts are filled with love songs, images of love, and the love of exploration has motivated people throughout time to invent and discover the mysteries of this

Earth. The shadow side of human love is that it often ends between people. The Toltec knew that the assemblage point was flexible enough to be dislodged into both *seeing* and also toward the feeling of love or loss of love by an unusual occurrence. It was easy for them to *see* the results or by-products of love, such as a kiss or a hug, but they *wished* to *see* beyond the emotional reaction. The illusive substance of love that linked the entire globe turned out to be awareness. When awareness reached a clear and pure form, even momentarily, it was named love.

Go to a crowded area of the city and take a position where you can view the passing people. Silence your thoughts and intend to become an objective perceiver of people. Observe all the interactions that connect one person with another, or those which alienate people from one another. *See* the moment before two people come together. You are trying to *see* the positive energy that occurs when they recognize each other from a distance and are happy that the meeting is about to occur. *See* the unified moment when two strangers interact over a connection made as one of them walked their friendly dog. On the negative side, *see* if people retain anger when accidentally bumping into one another. *See* the anger that remains when a car accident is avoided and yet the drivers still maintain a lingering emotion long after the event.

Witness human interactions as a search for connections, both positive and negative, within their unified assemblage point. They are always interacting in predictable ways between the bookends of extreme joy on one side and the emotion of extreme anger on the other. In between are residual feelings called melancholy, sadness, happiness, or boredom, but all consistent within the human perimeters.

It is interesting that we can actually make assessments of people from a distance, before we validate our *seeing* through any personal interaction. The ability of a human to detect a friend or a foe, or even someone with whom they will have a deeper relationship way before the contact is consummated is a remnant of our survival skills from the beginning of time. In order to establish a cohesive tribe the Toltec used *seeing*. *Seeing* the abstract qualities of another person was mandatory for the development of their advanced civilization. Incorrect perception of a situation could mean the end of their civilization. The end of the MesoAmerican culture of the Aztecs began with their incorrect *seeing* of the strange and hostile visitors from Spain, wrongly interpreted as the returning family of the Quetzalcoatl lineage of the Toltec.

Your ability to understand a person's essence from a distance is perceptual magic. I instruct my apprentices to always use the first few minutes of *seeing* a person they hadn't *seen* for some time, as the best moment of true *seeing*. Those few seconds would be the only time available for assessing any changes in the true subtle appearance of their friend. After those first few moments, a familiarity sets in and you gloss over the initial *seeing* with old filters. After the first few seconds the person is *seen* as the Known, instead of the Unknown. The first phase of viewing your friend is when you view them from a distance and they remain part of the panorama of strangers. This is the only time you can assess them with pure *seeing*.

The second phase comes when you recognize them suddenly in the crowd, and begin structuring their image with personal history. The last phase is an assembling of their personal history with you, and those glosses of experience reshapes them into Known familiarity. In the last phase you can no longer assess changes to their physical appearance with any impartiality. If your friend had gained weight or aged, or the more subtle aspects of their energy have changed, it is no longer as apparent as it had been during the first phase of *seeing*. In order to understand the subtle differences in people, the perceptual magician stands outside of the Known. They train their *seeing* for pure understanding of the energies that make up all living things. Being able to determine the most rarified qualities of a human is the peak of *seeing* abilities. And being able to *see* when love between two people enhances both parties toward a lifetime of awareness is a gift bestowed by the mature perceptual magician.

Now intend to *see* someone who is outside of the general assemblage point of the passing crowd. If you are fortunate, one day, your intention will be answered and you will *see* this person in a crowd of strangers. When this person arrives, *see* how they walk and interact within the chaos of random moving people, there is something different about their movement, perhaps it is their grace, or their confidence. Their body language is elegant and assured, their mood is one of detachment and engaged sobriety at the same time. You like this person. You could *see* this person becoming a trusted friend.

You *see* that this person interacts with the negative and positive life situations as a warrior: patient, sweet, and with a degree of cunning and even ruthlessness. You *see* them as someone who can be trusted to carry out a task and they would be impeccable with their every word. You *see* that your connection with them is not based on ordinary assessments from knowing them at work or play or family events. You are able to understand them without any physical interactions or conversations. This is a clear assessment without hesitation, and it feels true without internal reservations. This person is removed from your reasonable framework. They have activated a mysterious property waiting inside, that we have named love.

If your "looking" has developed into clear and accurate *seeing*, then it is your awareness that matches their awareness, and this is called in the Toltec map the "meeting of the Cores". There is something overpowering in their effervescent being that can only be defined as awareness. This person personifies awareness by simply being themselves. They are sharing that energetic field with everyone they pass, without shielded reservations. *Seeing* in such an assured fashion is outside the realm of reason. Trust your intuitive sense, and when you are ready, your Voice of *Seeing* will validate your assessment. If the Voice of Seeing is silent, then your desires and projections have clouded your true seeing. If the Voice of Seeing gives you an affirmation, then you have just *seen* and experienced love as awareness.

Once upon a time, I was at a party. I was still in high school and everyone was there with the *wish* to meet a special someone to become a boyfriend or girlfriend. There was hormonal love in the air. As I sat against the far wall observing the entire spectacle, I began to *see* humans for the very first time. One particular girl kept catching my attention through her ability to naturally attract the young men. It wasn't the way she was dressed or even her physical beauty, but a simple "licking of her lips" that seemed to hold the key to her success. Every boy that she approached was presented with the same subtle movement of her mouth. From my neophyte *seeing* her technique was startling.

I began an internal debate about what people did in an unconscious way with their bodies to form deep connections with other people. That night I left the party thinking that there was a part of human behavior that had been developed from the dawn of time; behaviors that attracted one person to another, with the smallest body language. I was *seeing* adolescent love as a primal bodily connection. This young girl was using a not-doing of lip-licking to accomplish her *wish* of attracting a boyfriend. It didn't matter to me if she was conscious of this not-doing, or a master of her seductive craft. Her actions had opened the world of *seeing* and this Unknown young woman had become my very first perceptual teacher.

II
The Path: Yellow Brick Road

The Toltec were artists of perception, who translated their *seeing* into monuments of stone. It was their intention that their pyramid temples exist for as long as the star worlds rotated above and for as long as the Earth world existed below. The great road at the center of the Toltec city of Teotihuacan led to the temple of the sun. This path was symbolic of the trail toward the white light of complete awareness that began with the browns and greens of the natural Earth and ended with the apprentice's absorption into the bright light of the sun. They painted their temples in polychrome colors and designed glittering mica murals that told the stories of their adventures in *seeing*. The reproduction of color to decorate their monuments and murals began with mixing ochre and red Earth tones and eventually they were able to duplicate the colors of blue and purple, the last and most difficult.

When they were training someone to *see*, the teacher first introduced the apprentice to the colors of the planet. They would walk the apprentice into nature and reveal the importance of the magnificent plumage of the quetzal bird, and the iridescent changes of color on a hummingbird. They would return from their walks to carve images in stone, and then paint over the rock with the brightest colors gathered from minerals and flower pollen mixed with tree resin to replicate all the hues of nature. Each color held power. For the Toltec, colors were connected to the way they perceived a friend or how a dangerous animal could camouflage itself in the forest. Colors were a path to understanding the designs of the universe.

The Toltec considered themselves to be "gazers". *Gazing* was the intended process of collecting valuable information with still eyes. This prolonged and contemplative *seeing* was a not-doing to unravel the mysteries of the Known and Unknown worlds. At the beginning of their apprenticeship, a candidate was given a specific color to use as a *gazing* tool. The teacher handed them a red flower, a green stone or a black piece of obsidian to study until the student understood all the nuances of that color and later, how to duplicate the color in their arts. The apprentice developed a life map from these explorations, particular and unique to their *seeing*, and a code to who they were, and what the future held for them personally. Particular colors became associated with the internal moods and predilections of each apprentice, and they wore colored amulets that portrayed their disposition.

Reading the colors of a person's luminous body was a skill of the accomplished perceptual magician. The transparent cocoon that encases every human and its variations from a milky white overlay to a beautiful shade of amber were *seen* by the Toltec magicians, and this information was used in determining the health and well-being of tribal members.

Shadows fascinated the Toltec, because these illusive filters were not made of easily classified colors. They could *see* the shadow's transformational ability to change all colors when it passed over them. Shadows were a common occurrence for all objects on the planet. All people had a transparent shadow image of themselves on the outside. The adaptable and abstract nature of the cast shadow reminded the perceptual magician of their own mysterious ability to touch all objects with their *seeing*.

The Toltec used shadows for the not-doing of *gazing*. When the apprentice was instructed to *gaze* at shadows, they were being led into a relaxed method of viewing the world for an extended period of time that put them on the magical road to *seeing* the essence of the Unknown world. The eyes were not forced into the "doing" of looking, but could relax their focus and enter a shadow space between wakefulness and dreaming. In this middle world of the shadow, the apprentice activated their imagination and their normal internal dialogue was replaced with internal silence, and eventually the clear Voice of *Seeing*.

What the apprentice understood through their *gazing* was that each shadow was an abstract representation of the spirit of the object that was casting the shadow, whether stone or human. Since the shadows were the opposite of the physical object of origin and were not solid, and yet they were made of the same dark properties no matter the origin, they became a window into the mysterious side of the planet. The shadow was *seen* as not only the other side of light illuminating an object, it also represented the abstract side of being a human; darkness was the mysterious unfathomable nature of life. When the Toltec looked into the night sky, they saw that darkness was the shadow condition of the universe.

For these perceptual magicians, dark and light were the predominate method of classifying the energetic duality of this world. Each color was balanced within a tendency toward either the darker or the lighter hues. Every person followed a path toward darkness or toward the light. The Toltec divided their mythology into the darkness of the underworld, the transitional grey of this middle world, and the brightness of the upper world.

All the worlds were connected with a transparent blue energetic column that could be used for inter-world travel with their Energy Bodies in dreaming. The blue energy of the connecting worlds column had the same properties as the small central blue core of every human. The column within a human's luminous shell could be *seen* as a transparent blue blush extending from the collarbone to the coccyx. For a Toltec magician the energetic similarities between the human core and the inter-world column signified that both were mirrors for one another. When the apprentice understood the symbiotic relationship between themselves and the larger sphere, they became accomplished transmitters that opened the perceptual possibilities for all others in their energetic field.

....

You are prepared to enter the next world with your young daughter. You are leaving the cityscape for the wilds of nature in the high desert. As your feet touch the soft white sand there is a feeling of anticipation, flowing up the path and through the transparent wall that separates the world of looking from the world of *seeing*.

The landscape is made of beautiful large beige boulders sculpted into rounded peaks or odd shapes that defy gravity and stay balanced on top of one another. Shiny olive green creosote bushes and powdered green sage, perfectly arranged by nature, line each side of your path.

You continue to walk into the higher portions of a magnificent canyon. The steep sidewalls are lined with immense boulders sculpted by wind and rain to resemble mythic giants or strange elongated totems. You hear the Voice of *Seeing*:

"The world of reason can be momentarily suspended and replaced with the world of seeing. The world of seeing is a wonderland of infinite possibilities that are activated with the intention to shift your frame of reference."

Your daughter points to a rock formation that looks like a camel face and observes that everything has its own special color. You are the encouraging guide for your daughter, and you use the moment to suggest making an intention for this particular walk. You recommend with a whisper that she might make a *wish*, and she says out loud:

"I *wish* to follow colors."

Walking further into the canyon you begin collecting color clues from the landscape. Continuing down the path you *see* many shades of the color brown, from the dark cracks in the boulders to the soft auburn cream color of the dried grass, but nothing significant seems to meet your daughter's approval. Then suddenly a brighter color catches her attention. It is a small red flower amidst all the muted brown tones that someone named a long time ago, the Indian Paint Brush. Your daughter announces:

"We will follow red."

You proceed along the dry sand river looking for some other feature on the Earth that is red while your daughter is convinced that a red Ladybug is watching from her leafy tower. Then you hear a high-pitched whistle sound. The sound is not Earth bound but is coming from the sky. You look up to observe a type of raptor bird flying high above named the Red Tailed Hawk because its tail feathers glow red when the sun is filtered through them. Its colored tail feathers are surrounded by the next color you will follow. The hawk is surrounded by the bright blue sky. Your daughter announces the next color in her color wheel:

"Its red tail has led us to the next color… blue."

The red-tailed hawk circles impeccably within the cerulean blue sky *seeing* all movement in its domain. The blue sky is made all the brighter with the contrasting redness of its tail feathers. The next color in your color-wheel walk has captured your attention, and you begin to follow "blue" as you progress along your path. You think to yourself about all the things in nature that might be the color blue; berries on a bush, the blue of moss, the glint of a Blue Jay feather. But nothing shows up, so you turn to tell your daughter that perhaps she should give up looking for something blue, because she may have reached the end of her color wheel path. But then something magical happens when you turn to look at her, and she exclaims:

"Your eyes are blue!"

You blink a few times and realize that your daughter is intent on fulfilling her beginning *wish* and continues onward to discover more colors; she is also a robust guide with abundant youthful enthusiasm. She has inspired you to proceed and you lift a sprig of green that you twisted off from a nearby bush that is silhouetted by the blue sky. Her enthusiasm is infectious and you laugh together at not *seeing* the unlimited possibilities of color in the landscape. The entire desert is dotted with unending "green" chaparral. You are living on a green planet.

You continue on the path, following the well-worn tracks left by a family of White-tail Deer. You discuss how people have named animals and plants after "colors": the Brown Pelican, the Green Heron, Pink Dolphins, the Red Rose and the Blue Hibiscus. The trail curves back and forth up the mountainside, because the deer know that crisscrossing is much easier then going straight up the hill. You climb to the highest point, a flat overlook that invites contemplation of the entire mountain range. Someone has been there before you, and left a beautiful circle of placed rocks with a space at the center, large enough for both of you to sit comfortably. This protective perimeter instills silent contemplation.

A larger stone with a different color has been placed along the radius at each of the four compass directions. The northern stone is made of black obsidian, an ancient remnant of the volcanic nature of this area. The southern stone is warm sandstone, while the east contains a translucent piece of granite, and the west holds a reddish rock from the canyon wall. You both marvel at the intention of the artists who left this beautiful arrangement for others to find in the future. The day is ending and long dark shadows extend from every corner stone across the white sand. The obsidian rock blends perfectly with its dark cast shadow. Your daughter asks:

"Are shadows colors?"

You suggest that your daughter *see* deeply into the shadows for an answer to her question. You both get down, very close, on your hands and knees and peer into one of the long shadows. The shadow is not black, but a transparent filter stretching across the white sand. Inside the shadow is a separate world. It is a place of day becoming night. It is the miniature Kingdom of the Shadow World. There is even a small green lizard hiding in the rock shadow from the bright sun. You *see* depth inside the center of the shadow, and then you imagine that the shadow is a lake of dark water in the midst of the dry desert.

Suddenly a strange odd shaped shadow flows across the ground. It crosses the bright sand, and momentarily flickers over you and then passes on across the canyon. Then more shadows appear on the sand circling the perimeter of the stone compass, casting shapes over you and your daughter. You both look up and *see* a group of black dots high above you; the source of the shadows. You *see* them as drifting shadow islands in a blue sea. As the irregular dots spiral downward, you *see* them gently transform into a group of circling black Condor birds. They are sky walkers following their own path in the sky. Their path is a spiral of invisible colorless warm air taking them higher and higher into the Kingdom of the Clouds.

A delicate rain begins to fall. The drops change the color of the rocks, and activate the dried moss on the larger boulders into bright green and amber hues. Your path begins to sparkle everywhere when the small mica chips scattered throughout the sand are washed clean. Then your daughter *sees* something new and points back up to the sky. It is an omen of great encouragement to continue learning the art of *seeing* and an appropriate ending for this day of the color wheel walk. Your daughter exclaims:

"A rainbow… All the colors!"

Not-Doing: Following Colors

Begin walking through the wild landscape or the city observing all the colors around you; the colors of the trees and flowers, the colors of the buildings and those colors worn by people. *Wish* this day that "silent knowledge" is held in the world of colors and that you will be able to *see* that knowledge. If you are guiding your child, *see* it as the moment you will introduce them to not only their favorite colors, but also to how colors can be used as *seeing* gifts from our planet.

The first "color" that appeals to you is the first color on your *seeing* "color wheel." If you are in wild nature, then you will be following the organic colors of the flora and the fauna of the Earth. If you are walking in a city, you will be following a combination of man-made colored products, like clothing or parked automobiles or building facades, as well as the occasional natural landscaping that appears during your walk.

Follow that *first color* that appeals to your temperament. This is a not-doing. It is the beginning of your "color wheel" and only at the completion of your walk will you begin assembling meaning in the world of reason. As you continue your walk, look for anything in the wild landscape or on the city streets that is in the general hue of that *first color*. When you encounter that same color or a color in the general range of that first color, walk over to that location. Remember that the *colors* are your guides. You change the direction of your walk when you *see* the next color clue. If you are following the color "blue" then go to the next location that shows up with the "blue" color appearing on something or someone.

Continue to develop your path by following this *first color*, no matter where it leads you. If you are following colors in the city, you will have a wider range of choices, from the clothes people are wearing, to the objects on the street or the colors in store windows. If you are walking in wild nature, then your color guides will be the flowers, animals, and elemental colors that you follow.

At some point, another *second color* will appear next to your *first color*. It is a completely new *second color*. You decide on the second color by feeling that its arrival on your path is appropriate in timing and place. You may have already begun associating the first color with events in your life when that color was predominant. In the early stages of a child's life, they are often asked to name their favorite color. They may pick a color that is the current fashion among their peers, or pick a color that few have chosen. In either case, it is a moment that will last a lifetime. It is often the beginning of a personalized life map, and their color choice will remain with them forever. When the *second color* appears, it begins the spontaneous ordering of events, that is now occurring through not-doing as you progress along your walk. You are beginning to *see* a pattern within your color wheel; your color choices activate associations with events in your life that can be recapitulated at the end of the walk.

Follow the *second color* as it leads you along your developing path. Each time this *second color* is seen in the landscape or on someone or something in the city, go to that location or turn toward that new direction. This mapping of your route, using color, is developing your internal map of understanding the role of color in your life. Colors have taken on aspects of both the emotions and/or the elements; red often represents passion or fire, while blue is associated with the cold or an unavailable persona. Each world culture has prescribed multiple meanings to colors. The Toltec considered green as a sign of instability, while blue was the color of pure energy. When the color red was *seen* on the luminous cocoon or the interior core of the human body, it was a sign that the person needed a healing. White or a shade of amber was the color of the healthy luminous shell.

Continue on your magical path to a third and fourth or more colors until you have exhausted your personal color wheel. Just as the beginning of your walk was a not-doing, so also the end of your walk is a not-doing. There is no formulated ending, but only a feeling that you have collected a sufficient amount of information from the colors to be energetically complete.

Sit and reflect on your choice of colors and how they are connected to aspects of your personal life. *See* how colors affect your mood and change your environment. Think of your favorite color and review why it is your personal guiding color and where you first encountered it during this walk. Reflect on the order in which the colors appeared. Ask yourself if the order of the colors showed a gradient toward bolder or subtler hues. Following color is a not-doing until you recapitulate your color wheel at the end of your walk. The review happens when you make order out of the not-doing with your "doing" reason. You now have structured knowledge about yourself through a not-doing. The "doing" comes about at the end of your walk in a dedicated contemplation and ordering of the entire color walk. By allowing "chance" to enter into your decision-making process, you are forming a creative map toward personal awareness through not-doing.

When I was a young apprentice I used a "color-wheel map" to navigate my walk in Golden Gate Park in San Francisco. Although I was just beginning my perceptual magician path, I was already determined to be a visual artist. Color was in my painting palette and its study was part of my artistic practice. I questioned my entire range of *seeing*, and would often close my eyes as I walked through an area, just to observe the outside shadow and light change the inside color behind my closed eyelids.

This day in the city park I was depressed about my chosen profession and wondered if I had the necessary fortitude to be an artist. I intended to follow the guidance of color and began with the darkest shade of black I could locate. To me the darkness represented my current state of mind. As I *gazed* into some black moss at the bottom of a forest pool, I thought about my chosen artistic path, and if it was too difficult a life to pursue. As I sank deeper into my own darkness, I couldn't help but eventually notice all the Water Lilies floating on the surface of the pond. I wondered how I could have missed them before, and decided to rise out of the blackness.

When I moved my vision upward to the Water Lily pads I felt saved from depression by those green floating islands. The pinks and reds of their flowers brought me great happiness. The purple flowers made me feel even more connected to the possibilities that I would succeed in my profession. It was at that moment that wind shook the nearby Dogwood tree and sent a current of white flower peddles cascading onto my body. I suddenly felt not only happy... but also enlightened. Color had not only brought me back to my best self, it had orchestrated a map that matched the exact internal emotions I was experiencing. That day I walked out of the park confident that I would be a great artist.

Not-Doing: Shadow Gazing

Begin your walk in wild nature at the time of day when the shadows are dark and long; in early morning or at twilight when the sun is setting. Scan the terrain until you find a particular shadow that covers a large area or is particularly dark, or a shadow with unusual contours. It will be a shadow that you enjoy spending an extended period of time observing. It could be the random moving patterns of shadow leaves or the stable shadow of a large boulder. Sit just outside the border of your chosen shadow.

Observe the boundary of the shadow, where the dark edge stops against the bright non-shadow area. If you are shadow *gazing* into a moving field of shadow leaves cast by a tree, their fluctuations will defuse the pattern. *Gaze* intently into that meeting edge between the light and the dark for a few minutes. Let your eyes relax. Let them go slightly out of focus. Go beyond your normal time frame of looking at an object. This extended time of looking will eventually open the possibility of *seeing*. *See* a subtle vibration at that meeting edge between the shadow and the bright area outside. The longer you *gaze* the better the chances of stopping your normal internal dialogue. That moment of internal silence opens a space for the Voice of *Seeing*. The Voice of *Seeing* will poetically remind you of your own life with shadows. Perhaps dark places scare you, or you find comfort in the warmth of the sun and avoid standing in shadows because they activate a feeling of aloneness. Pay attention and let the Voice of *Seeing* help you remember the first times you interacted with shadows in the outside world or in your interior world.

Next *still* your vision for a few minutes in the central area of the shadow. *See* if the shadow has a lighter cast at the center then at the edges. Mark with a stick or small rock the edge of the shadow, and then return later in the day to *see* how far the shadow has moved from its original position. Observe the changes in your own shadow as it passes over irregular surfaces. Contemplate upon the duality of planet Earth; how most of our decisions can be categorized as either dark or light in substance or metaphor.

For the Toltec, *gazing* at shadows became an exercise related to time, and the movement of the shadows during the day, a way to count the cycle of the year. The movement of shadows inspired the designers of the pyramid in Chichen Itza, Mexico. As a tribute to the Quetzalcoatls, the lineage of the Feathered Serpent perceptual magicians, the stone steps leading up to the top of the pyramid cast a moving shadow of a serpent during the spring equinox. The magical appearance of the snake shadow due to the precise arrangement of the stones was a *gazing* exercise for the entire population. These perceptual magicians had united the sun with the solid rocks of the Earth in one fantastic work of perceptual art. The entire tribe would gather at the time of the equinox to patiently *gaze* at the magical appearance of the snake shadow and its descent down to the Earth.

The Quetzalcoatls were the master magicians of *seeing* perception in MesoAmerica. It was their influence that brought high culture in art and architecture to the region. From the use of arranged stones in a Personal Power Spot to the building of stone temples, these perceptual magicians infused their creations with *seeing* knowledge. It was their intention to move the participants away from just looking, and into another syntax that they named *seeing*.

Shadow *gazing* was just one part of a large body of color not-doings utilizing natural phenomena. There were elemental *gazing* techniques involved with water in all its forms, from lakes to rain to fog. The transparent nature of water, and its ability to alter the color of everything it saturated was of particular interest. The use of fire *gazing* was a profound method taught around the community fire. The apprentice eventually used the changing color around the flames to shut off their internal dialogue and imagine worlds of great intensity. All the animals and their colors were observed closely to utilize them as *gazing* vehicles, revealing perceptual abilities to vanish into the landscape with their perfectly matched bodies. Plants with their varieties of colors and textures became *gazing* tools.

As a Toltec Father, I took every opportunity to guide my young children into the map of *seeing*. The gaming board of the wild forest trained their eyes into the subtle realm with unlimited hues of plant and stone. The second phase of the teaching involved each child taking a turn, running ahead of the rest of the group to define the path by placing found sticks, flowers or colored feathers along the path to define the direction for the upcoming group. Eventually the directional markers became more and more subtle, and only with intended *gazing* could the proper marker be detected.

My final not-doing in the realm of leaf *gazing* was to paint a stack of dried Fall leaves in various shades of black and white and grey. I went early down the path and left these painted leaves at various junctures. It was the Fall season, so many leaves were on the trail in shades of brown, red, and yellow; only my leaves were in the shades of black and white. My painted leaves were disguised in plain sight, but invisible without moving your attention out of the Known Fall colors. Although the colors of leaves on the trail were in extreme contrast to my black and white shaded leaves, the chaos of leaves still demanded *gazing* concentration. I still remember the sound of my three young children's voices, and the crunching sound of breaking dried leaves, knowing that they were coming closer as I sat at the end of my colored leaf trail. This moment of quiet anticipation made me feel young again. As I sat alone in the forest I moved back in the Wheel of Time to memories of myself as a child playing a game of hide and seek with others, hiding in a perfect spot where no one would ever find me, but this time in reverse: Instead of hiding, I anticipated being found by my children and the joy of the resolution of our not-doing. With this not-doing I was not only their Toltec Father, but I was also one of their best friends.

III
The Wish: Upon A Star

The Toltec realized that one's intention, at the beginning of any walk, had to be voiced out loud. They instructed the apprentice to arrive at any session with one clear sentence that defined their *wish* for that particular walk. The apprentice would stand at the threshold of the path and pronounce their *wish* before the elements, before all the creatures on the landscape, and to the spirit, believing that if their intention was accepted, then everything in the Power Spot would contribute to the success of the walk.

They believed that without verbal clarity and a defined *wish*, there could be no clear answer. The answer would come in the form of omens from all of nature's participants: the plants, the animals, the sound of the wind, and eventually from their internal Voice of *Seeing*. With this larger connection, everything was possible for the perceptual magician. They relied on their connection with a greater power, outside of themselves, to accomplish feats of magic, and they called this power, the spirit. The spirit resided inside all things, and the emanations of the spirit flowed through, animating all life.

The beginning of any larger understanding of the world meant overcoming self-importance, the belief that the apprentice was the center of the world. The teacher paid attention to the type of *wish* voiced by the apprentice. Their *wish* revealed their interior patterns; learned behavior that offered clues to the apprentice's internal dialogue.

Sometimes in voicing their *wish* a hidden personal fear would be revealed. The teacher wasn't interested in a minor fear, but wanted to *see* a deep fear, a fear that touched on the feeling of death. The Toltec believed that deep fear was a tangible emotion that could be intentionally activated for the purpose of learning how to make that fear into a lifelong friend. A fear that touched on the feeling of death was not an enemy to be eliminated. This fear, once located, could be used as a catalyst for deep personal growth.

As dedicated researchers, they would place each of their apprentices in their personal precipice of fear; this could be on the edge of a steep mesa, or alone in the darkness of a deep cave. To deny this feeling was not possible for a perceptual magician. Death was the final gate for every living thing, and the Toltec method of approaching death was of ultimate importance. From the very first walk, the teacher was leading the apprentice to a realization that they were eventually going to die, and that death mirrored deep fear at its core. They believed that this one great fear that touched on the deep feeling of death was a friend that walked by your side till the very end. Death as a real presence was a true and lasting Voice of *Seeing* that remained a dedicated and impartial witness to all the events in their lives.

For the Toltec, the idea that they could form a bridge from the Known world to the Unknown world was a challenge worth pursuing. The Known world consumed most daily activities, such as survival for food and shelter, while the Unknown world revealed itself in dreaming. Dreaming could occur with open eyes or with closed eyes. Some of the first *seeing* adventures of the perceptual magicians involved simple eye not-doings, like squinting or crossing their eyes. Later more elaborate methods were devised such as having a small bead suspended on a string that dangled in front of their eyes. A stick was attached to the top of the apprentice's head, so that the bead was hanging down parallel with the eyes. The apprentice was instructed to *gaze* at the bead until their eyes crossed. Walking through the landscape with this contraption on their head forced the student to *see* two overlapping realities. The varieties of hanging objects varied from simple stones to shiny crystals that reflected the light of the sun. They considered crossed eyes to be a symbol not only of a practiced perceptual magician, but also as a sign of great beauty.

The not-doings of the Toltec evolved from simple devices to more elaborate situations. They developed methods that suspended the apprentice off the ground with ropes, or lowered them into deep caverns inside the Earth. All these not-doings were arranged with impeccable care for the safety of the apprentice, but allowed the apprentice to feel the intensity of aloneness, and as a consequence, the exultation of survival after the exercise.

Tricking reason was a worthwhile pursuit that rearranged the apprentice's accepted mode of thinking with intended creativity. In this process of altering the Known, the apprentice was faced with a clear choice. They could continue believing in one frame of reference or they could enter into another world of perception; one that offered unlimited creativity.

....

You have reached a boundary. This new territory is littered with the personal history of your civilization. Your daughter is also at the boundary between childhood and adolescence. Together, you step into the transition.

You are walking with your daughter over this boundary. You traverse out of the city park on your way to a place where the wild trees begin to sway and the rivers are still free. When the urban environment begins to encroach on the natural outback, there is sometimes a pile of manmade artifacts left at the area between the manmade and the truly wild world. It is the place where two worlds collide.

You have reached an outback dump; a place where people unconsciously discard the remnants of their temporal life. These are the artifacts of your civilization, left at the threshold between the tamed manmade and wild nature. You look around at torn sofas, cradling stuffing clouds, and car frames overgrown with weeds, half buried in the ground like rusty dinosaur bones. The ground you are walking upon sparkles with irregular chards of tiny broken glass pieces from house windows, scattered rectangular safety glass, and the colored glass of broken bottles.

These walks are power walks. They empower your daughter with the courage to interact with the mystery of new places. She is learning to make an intention when she faces these new worlds, to make a *wish* when she enters into these unique classrooms. She is turning a challenge into a creative interaction. Your daughter makes a *wish*:

"I *wish* that we will find the Crystal City."

The soft green field suddenly abuts a hardscape of broken glass. You step across the threshold into another world, a place of disturbed Earth. But with your daughter's *wish* this ugly environment is transformed and your looking becomes *seeing*. You *see* the landscape of broken glass become a field of fine jewels spreading out into infinity. And the Voice of *Seeing* says:

"The way of seeing is paved with the intention to see. Without you intending to go into the mystery, there will be no mystery. Leading an apprentice into their imagination is leading them into seeing. Building worlds is the highest achievement of the perceptual magician."

And then your daughter *sees* the fulfillment of her initial *wish* and names this place, the Crystal Kingdom. She adds:

> "You are the King and I am the Queen of the Crystal Kingdom."

She waves her hand across the landscape, rearranging the molecules of creativity, transforming all that lay before her. Your daughter has also transformed your continuum. You have become a King. You balance a rusted rim of a small bicycle on your head as a crown, and ceremoniously give a bent windshield wiper to your Queen for her scepter. You both walk through the glittering grounds of your Kingdom that has been transformed from an eyesore into a treasure trove. Then you sit upon two stacks of discarded automobile tires that have become cushioned thrones to observe your Kingdom. And then the Voice of *Seeing* says:

"Seeing is the ability to rearrange your Known world. You have been taught since birth to accept the names of things in order to reach a consensus with your fellow humans. Through seeing you stop the classification of the Known and release a wave of hitherto untapped energy."

Your Kingdom is a wonderland floating on clouds of twinkling stars in the amber setting sun. The old car parts suddenly reveal their secondary nature and drift through the star field like uncharted islands with castle spires and misty mountain silhouettes. You sit in the castle ballroom. And then your daughter says:

"It is night now and these are all the stars."

She has transformed the ugliness of manmade waste into a beautiful world; she has transformed a garbage dump into a Crystal Kingdom drifting in the night sky with her *wish*. This is more then a child's play; this is the process of creatively moving the human assemblage point into a different syntax of interpretation. Then your daughter points to a figure on the side of the ballroom and exclaims:

"We have a visitor to our Kingdom!"

At the far end of the glittering ballroom is a bear. He is standing completely still on all four legs. You sense that he had silently entered into your domain and that he is not afraid. But you and your daughter look at each other with both delight and fear. You stay still as a bear. The only movement in the entire ballroom is the wind ruffling his shaggy coat. After a few minutes you get up and cautiously walk toward the bear. You are nervous about approaching a wild animal, but you are intrigued by the possibility of verifying this encounter with a closer inspection. You begin an internal debate about fear of wild animals and what is a safe distance to maintain. Just then, you hear the Voice of *Seeing* say:

"You are seeing your personal fear. See how it touches your internal core. Our fears are vehicles for transformation. Often they are our greatest and truest teachers to the very end. They are also tricksters. See as you move closer to those situations which activate your fear, how the feeling of terror can suddenly transform into joy".

There is no safety zone for this bear. He remains completely still, until you are only a few feet away. But you had to get this close in the dim light to realize that the bear is not a bear; he is a collection of old clothes stacked on top of a rusted shopping cart. As you walk back to your daughter seated on her tire throne, you reflect on *seeing* your fear. You reflect on everything in the world that emphasizes fear and how this limits and often defines our perception of reality. Your daughter is becoming a young woman now, prepared to face a world of real dangers amidst the wonders. When you reach your daughter to share your "bear revelation", she has already *seen* the outcome of your adventure, when she says:

"Everything in the world helps you *see* more of yourself."

Not-Doing: Setting Intention

Walk to a transitional area; an area where the city meets the country. It must be a forgotten area with discarded artifacts that have reached the end of their usefulness. These are areas that were once part of your world: abandoned factories, dilapidated houses, or makeshift dumping grounds.

Say your intention *wish* out loud. Your *wish* should be concerned with *seeing* ways of transforming your personal world from a waste dump into glittering clarity. It is a *wish* for transformation. Look around your chosen environment and *see* if you are drawn to a manmade object that can be used as the first artifact to transform this contemporary dump into a wonderland of the imagination. Remember, this is a not-doing, so your choice of objects is based on instinct: your first impulse, retaining the wonder of its sudden appearance.

Go over and pick up the small object and hold it in your right hand. This intuitive selection will now represent what is "wrong" and in need of correction in your personal world. As you investigate the object, pay attention to your budding Voice of *Seeing*. Although it may still seem distant, this internal whisper is something that has been counseling you since birth. You were not acknowledging the extent of this advice, nor the origin of your decision-making. The Voice of *Seeing* has always been counseling you on things that weren't working in your own life, but you replaced its clarity with your own hesitation and confusion filters.

Now pick up another manmade object. This "second object" will represent the exact opposite of the first artifact. The second small object represents everything that is "right" in your personal world. Hold this object in your left hand. Consider that each hand now holds a representation of the duality in your world: night and day, anger and love, bitter and sweet, or even life and death. Your left hand holds the object that represents what is working effortlessly in your life. Your right hand holds the artifact that represents what is clearly not working in your life.

The Toltec divided the world into two parts: the "Tonal" which is the place of the Known, and the "Nagual" which is the place of the Unknown. You can also *see* the object in your left hand as your Nagual, while the right hand holds the Tonal. In the right hand is the world of everyday life, with all its difficulties and obstacles. Your left hand holds the world of your dreams and the abstract nature of the universe.

Arrange your two gifts into an altar in the midst of all the surrounding chaos. The altar is a tribute to both worlds that you inhabit: the world of suffering and confusion and the world of joyful mystery. Stand back and *see* your altar and its placement in the landscape of manmade clutter as a floating island of sobriety. You are *seeing* yourself as represented by your personal altar in the larger world of manmade chaos. *See* how the objects you selected for your altar, and the matter in which you assembled your altar answers your original *wish* for transformation. You have transformed discarded objects into intended artifacts of counsel. *See* how you can also transform the chaos of your life into a beautifully arranged altar representing your awareness amidst all the refuse of unconsciousness. You have also begun to use the Voice of *Seeing* to assemble clear answers for a life transformed.

Not Doing: Tricking Reason

Go to a transitional area that has now turned into a dump of manmade objects scattered across the previously wild terrain. It can be an abandoned factory, an overgrown derelict development, or any place that represents the meeting of deteriorating human presence being reclaimed by nature. Sit at a distance from this wasteland, sufficiently removed, to *see* a large section of the area. Scan the grounds until you find a discarded pile of objects that presents an unlimited amount of visual possibilities. You will use your *seeing* to transform this assemblage of objects from the discarded Known into the resurrected Unknown.

This not-doing is rather simple, and most people do this exercise spontaneously when they look at a mountainside; they often *see* images from their imagination in the complex formations of rocks. People naturally make order out of chaos and easily imagine familiar images they associate with their experiences from the abstract arrangements of stones; shapes that remind them of visual memories from their life collection.

The most familiar not-doing for most people is cloud *gazing*. When you *see* animals or representations of the Known world in the abstract shadow and light patterns of clouds, you are performing a *gazing* not-doing. Even the depiction of zodiac figures from random stars groupings, by aligning them with straight lines into animals or figures from mythology, represents another way humans use *seeing* to humanize and make sense of abstract nature.

Gaze from a distance at your selected pile of objects with *half closed* eyes. Your view should be relaxed and slightly out of focus. *Gazing* does not involve staring. It is a relaxed method of viewing. Imagine yourself having the patient eyes of a cat waiting for a mouse to enter the scene. The cat is not focused on any one thing, but is paying vigilant attention to the whole area for any changes. The Toltec recommended that their apprentices squint or blink their eyes rapidly when transforming the Known into the Unknown. By squinting their eyes, the apprentice was placing a filter of fog over the outside image. Their eyelashes could also be used to make fine webs of light to filter and distort their normal view. By blinking rapidly the apprentice was dividing the outside world into separate frames of still pictures.

Both not-doings easily changed the apprentice's normal viewing pattern and usual internal category making. These not-doings removed internal dialogue using a physical action that pulled the apprentice's concentration away from normal patterns and thereby momentarily short-circuited reason.

Concentrate with *squinted* eyes on this assemblage of your choice until it transforms into something different from your first completely open-eyed viewing. Your view is being replaced with a new set of images that are being creatively activated by working with your perceptual capabilities. The pile of refuse can change into an animal, or an inspired contemporary sculpture; but it is no longer just a pile of discards within the landscape.

Sharing your discovery with another person is often an interesting dilemma that points out the unlimited variety available in a single grouping. The apprentice may *see* something completely different in the exact same assemblage. The teacher should never correct the apprentice's *seeing*, and particularly not a child's interpretation. There is no right or wrong way to view the assemblage. *Seeing* is the activation of unlimited creativity.

I have often been amused when pointing at a distant hillside pile of rocks that looks like an image of a giraffe to me, and asked the apprentice if they *see* the same thing. More times then not, they will shake their heads in the affirmative. When I ask them to elaborate further on what they are viewing, it turns out that they are *seeing* the giraffe in a pile of rocks that is in the foreground to the one in which I was referring. We all see the arrangements in nature from a myriad of perspectives. The fact that a giraffe can be hiding anywhere in the thousands of rock forms is a validation of the unlimited creativity of each participant in the adventure of *seeing*.

Not-Doing: Seeing Your Personal Fear

The act of creatively *seeing* into an arrangement of inanimate objects can be used to understand your personal fears. Every apprentice harbors fears that manifest in unique ways as you walk through the landscape. Some people are afraid of wild places, some imagine a rattlesnake around the next bend on the path, or perhaps they face a personal fear of heights as the trail winds along a steep rock face. Fears in the city are numerous, especially on dark streets. Most of these fears are realistic and some fanciful, but all are connected with real emotions that can be confronted in a perceptual manner.

As a Toltec guide, I eventually lead the apprentice toward a revelation about the one great fear that is connected to the feeling of their death. This is not a minor fear. And there is only one great fear for every person custom tailored for their being by the Spirit. It is not unusual for apprentices to *wish* for a resolution to this one great fear in their life as the intention *wish* for that particular walk.

Seeing certain fears in your children often mirrors your own personal fears. As a Toltec father, I had made my own fear into an Ally by the time my children had to face their fears. By *seeing* the makeup of your own fears when they appear in your children, as a father guide, you can end the energy wasting dominance of a particular pattern engrained in your entire lineage. As a perceptual magician you can intend to end a dominant trait that limits the awareness in your children, and in turn in all future generations. Perhaps you will be able to trace this particular fear back to your own parents. Maybe a story will be dislodged from your memory that was told to you by your mother about her fear of drowning. You can *see* how this fear prevented her from ever enjoying swimming. Your intention involves making sure this trait is not passed on to your own offspring. With your *wish* and the follow up not-doings you are helping the future generations into clarity about the natural joy inherit in swimming.

A particular lineage fear can be *seen* as a creative challenge that arises every time your children are placed in a difficult situation that activates this fear. Some lineages have a fear of heights, or a fear of being in a confined space, or of speaking in public to a large group, or a myriad of other personal fears that have prevented a full interaction with life on this planet. As a Toltec father guide, you can set up Play *not-doings* that confront the deep nature of these fears, and rearrange these limitations in your children, and thereby in future generations. Through proper *seeing* and intended *not-doing* practices, this trait can be transformational. If your lineage had consistently been afraid of heights for instance, and you *wish* to transform that fear within your children, you can arrange *not-doings* to confront this fear of heights, without placing your children in danger. Approach this very important not-doing as a Play. It is not talking therapy, but a magical not-doing using bodily senses, removed from mental gyrations. Remember that not-doings do not rely on explaining reactions with words to be effective. On the contrary, a successful not-doing cannot be logically explained.

Take your young child to the nearby playground. Name the area after a mythic place that encourages their internal warrior spirit. By naming the place, you are instantly transforming their regular playground into another world. If your child is afraid to go across the monkey bars, make that a bridge to escape the coming pack of wild creatures from another planet. Encourage their imagination, which in turn activates their *seeing* potential. *Wish* out-loud that your young child will cross a river of crocodiles in the sand below, when they go hand over hand across the parallel bars in the playground. By encouraging their heroic crossing, their fear is used as a motivating force; it is transformed into a personal friend, not a personal fear.

If your child is afraid of heights, the high slide in the playground can be transformed into another perceptual event, one that encourages magical *seeing*. Remember that these not-doings are custom tailored to your particular situation. Create the titles and story that is drawn directly from your child's interests. The slide can be named the Palace of the Princess, the Tower of Discovery or any other labeling that they can associate with their lives that will stimulate their *seeing*. By creatively guiding your child up the ladder to the Palace of the Princess, the climb becomes effortless and your child will be eager to accomplish this not-doing. In the process they will lose their old syntax and embody a mythic spirit that will last them a lifetime. Their "will" is activated. From a Toltec perspective, it is far more then building self-esteem. In the process of facing their fear in creative play, they are forming an internal map of *seeing* how a warrior confronts a world that has "fear" manifested in many guises.

As a Toltec father guide, introducing a young child to the idea of death is never necessary. Along their path, they will naturally encounter the death of a pet, and you probably will find yourself helping to bury a family dog or cat, or even a discovered wild animal. Later, they will most likely attend the death of an elderly grandparent. The idea of symbolically burying an adolescent idol is a ritual that my daughter and I did when one of her favorite music stars passed away. We used some of the symbols of his music and placed them inside a tree to put some closure on her spirit attachment to someone she only knew through his musical art.

Once upon a time, I was walking with my daughter in the wilds of Bryce Canyon in Utah. She was still very young, but during the walk I confided to her about death. I knew that she would not be attached to what I was saying, and would surely forget at her young age, but that moment and the breathtaking landscape took me with its power.

As we sat amidst the giant red amphitheater of erosion created hoodoos, I confided to her that one day I would be gone. When I died, I told her that I would fly over this exact spot and say goodbye to this beautiful Earth for one last time. My daughter was a silent witness that day within the canyon spectacle. Her child innocence was far removed from thoughts about the end of life. She was holding a small wild flower and performing an enchanting movement of gently touching the ground with the end of the flower and then placing the tip into the air. At first I thought that she was making a flower bridge by touching the Earth with one end and touching the sky on the other.

She repeated this gesture over and over with complete attention to her actions. I *gazed* with fascination until it became clear to me that she was not-doing. She was drawing a picture in the sand by dipping her flower brush into an invisible ink bottle suspended in the air. She was also punctuating the moment with a magical pass that would help me remember this day, until I did fly over Bryce Canyon with my Energy Body after I had died.

THE WISH: UPON A STAR

IV
The Powers: Super Heroes

The Toltec were deeply concerned with the idea of stopping the internal dialogue, those voices of both fashionable reason and dualistic debate that torment the mind of the undisciplined individual. Attempting to stop these distracting voices inside their heads that relied on Known solutions, involving indulgent behavior or further exterior conversations with others, was still in the domain of the island of reason. They inferred that the island of reason was only one half of a complete perceptual warrior.

The second half of their attention that needed to be studied was the realm of the not-doing, the activity that opened the creative source of the Unknown. In order to assist in stopping the constant chatter that left no room for creative *seeing*, they developed not-doing techniques that involved not only their eyes but also their other senses, like hearing, which they found could short circuit the dominance of normal looking connected with their internal dialogue.

They began their studies by taking the apprentice to nature, where no human voices or noisy village activities masked the purity of the sound of the wind. Instead of using the dominant sense of looking to begin, they had the apprentice shut their eyes in peaceful solitude. When they closed their eyes they eliminated the principal tool for looking at the Known, their open eyes, and could then use hearing as the primary vehicle of perception. Hearing the voice of the wind moving through the leaves, or focusing on the voices of the birds, the crickets and the frogs, freed them from the dominance of looking with its attachment to word classification and helped to stop internal dialogue. They found that the images inside their closed eyes became a creative journey to remote locations, carried by the pure sounds of nature.

Stopping the internal dialogue meant eventually being able to *see* into the nature of everything making sounds around them. The association with a particular image was greatly diminished when the apprentice used only sound stimulus. A bird sound could be dislocated from a particular bird shape and be used as an abstract carrier for dreaming. A chorus of frogs, the rustling of leaves on a bush, or the energetic movement of wind, could carry the dreamer into perceptual flights without the limitations of looking.

The Toltec developed specialized instruments that enhance the use of sound specifically for dreaming. Dreaming was the mysterious side of humans. Their practices with closed eyes opened the possibility of traveling during sleep. Dreaming is the not-doing of sleep. A teacher would sit next to the dreaming apprentice and play sound instruments that were designed to enhance the abilities of traveling the Unknown. These instruments were called *seeing catchers*. They were used in the day and the night to assist in dreaming. Sitting on a Power Spot they used the *seeing catcher* to produce an additional sound that seamlessly complimented the other sound players in nature; the crickets, the birds, the wind and the water.

Seeing catchers were originally made from objects easily found in nature: a straight leaf or blade of grass, placed between the lips and activated by breath, to produce a soft humming sound, or the simple rubbing of two stones together to mimic the sound of a digging rodent. As time progressed the *seeing catchers* became more elaborate: gourd rattles, cloth bellow accordions, and metal jaw harps. By participating in a not-doing with a *seeing catcher* the overall ambient sounds of nature revealed their harmonious presence, and in this process freed the apprentice to dream with non-human musicians.

Their explorations in dreaming led them to the conclusion that every person had two distinct modes of interaction with the world: the day-to-day physical body and what they called the Energy Body. The Energy Body was used in the world of dreaming. The Toltec would sit the apprentice on their Personal Power Spot and with closed eyes have them *wish* project their Energy Body out from the area in which they were sitting. Dislocating the Energy Body from the physical body was a not-doing that everyone did naturally in deep sleep. The teacher was simply reminding the apprentice of their ability by placing them in the right situation and having them make the proper intention.

The body wants to participate in its totality as a perceiver, and making the apprentice aware of using the *wish,* freed the body from the chains of the Known to use its full capabilities. The Energy Body was free to roam the landscape and observe anything of interest. The second phase of the instruction was to have the Energy Body detect something particularly appealing in the landscape, a special rock or perhaps a pool of water, anything that caught the attention of the eyes of the Energy Body.

The apprentice was then instructed to get up and walk with their physical body to the area they had just *seen* with their dreaming Energy Body, and locate the rock or the pool of water in the landscape with open eyes. This not-doing was practiced until the apprentice was able to feel their awareness in two places at the exact same time; they felt the physical body sitting on the Earth, while simultaneously they could *see* a remote location with their Energy Body. The crux of the exercise was to retain a feeling of being in two different places at the exact same time. When the apprentice could fully cognate the feeling of flying outside their physical body and still understand that they were sitting on the ground in a physical body, then the crack between two perceived worlds had opened. The apprentice had leaped back and forth riding the parallel lines of awareness available to perceptual explorers.

....

Your daughter is a young adult. She has reached a stage in her proper assimilation into the Known world, so that she is ready for the advanced studies in perceptual magic concerned with dreaming. You stand at the threshold of explorations with the Energy Body. Then you step forward.

You are hiking into a mountainous desert landscape with your daughter. As you climb higher into the mountains using a jagged rock passageway, numerous protruding ledges extend from shallow caves. You finally sit to rest under one of the canopies of rock that provides shade for you and your daughter. You share stories with her about the ancient Toltec who visited mountains that were inaccessible with the physical body, but were easily explored with their Energy Bodies. She listens intently about how a perceptual magician is given one or more *super powers* that enable them to accomplish incredible feats from the perspective of the Known. These powers are human abilities that they have magnified into super sensing talents through disciplined practice.

You instruct your daughter to close her eyes, shut off her internal dialogue and enter into "silent knowledge", the place where the Voice of *Seeing* resides.

The crickets, silent when you first entered, return to singing their songs in a unified chorus of call and response. The birds feel safe enough to be themselves and add high pitch chirps to the symphony. Then, approaching from far away is the sound of the wind moving through the chaparral. The wind reaches your Personal Power Spot and it is strong enough to pick up loose twigs and create the sound of rain when hitting against the top of the stone canopy. You say to your daughter:

> "We are now in the Kingdom of the Wind and the wind can move us anywhere we *wish*."

You let the sound of the wind surround you and lift your hearing upward within its funnel of intensity. Instruct your daughter to remember what the desert looks like behind her closed eyes. Guide her to remember all the plants and rocks she passed today on your walk to this rock ledge; the details of colors and shapes she collected with her *seeing*. Instruct her that the desert can be *seen* with her closed eyes if she once again feels herself gliding down the sandy path.

Then guide her to *see* her Energy Body lifting with the help of the wind. Tell her that her Energy Body is her other body she uses in dreaming and it has flying abilities. Guide her to a feeling of opening the center of her physical body to release the Energy Body. This body is invisible, capable of traveling on the wind to any distant location. She can *see* anywhere and everything in the landscape with the eyes of her Energy Body. You instruct her to follow the sound of the wind as it pulls her Energy Body up and then over the desert landscape far below. And then your daughter says:

"Let's Go!"

Your daughter feels like she is a flying red tailed hawk chasing its prey across the vast desert panorama, *seeing* everything below in extreme clarity. You fly with your daughter over the land, keeping her Energy Body next to your own. She is safe within your energetic field to travel anywhere. And then you hear the Voice of *Seeing*:

"You are using your Energy Body to fly on the wings of perception out over the surrounding landscape, as far as you wish to travel. The only limit is your ability to feel when your intention has ended."

You glide over the desert arroyos with your daughter and her awakened flying abilities. You *see* the large boulders and scattered bushes far below and listen to the sound of the wind get louder as you go faster and faster, confident of your flying powers. The vastness of the landscape spreads out beneath your Energy Body and you even *see* your own strange shadow shapes moving across and over the ground below. You suddenly turn upward and head toward the bright light of the sun. You feel the more intense warmth on your face and then when the light is too bright you turn around and instruct your daughter to slowly descend back to your sitting spot under the cool rock ledge.

And then you hear the Voice of *Seeing*:

"The use of sound as a dreaming vehicle is a super power. The sound of the wind or the notes of a manmade instrument can capture attention and take the experienced traveler to distant lands. A magician can focus on natural sounds and augment them with his own manmade sounds. Hearing is also a method of seeing."

You take out a small child's accordion from your backpack. It is your personal *seeing catcher*. The sound from a *seeing catcher* is designed to stop internal dialogue by aligning with the crickets, the birds, and the wind. Your *seeing catcher* is an extension of your *wish* for spirit flight. The round handles of your personal *seeing catcher* fit perfectly inside each of your palms. You play the tiny accordion in a smooth slow cadence. The sound is one chord made from many discordant single notes, activated when you squeeze the small bellows back and forth. Although it is a small instrument, its sound fills the entire landscape.

The sound of the *seeing catcher* is amplified by the immense silence of this desert. Tall Yucca stalks on the far edges of your peripheral vision frame the panorama-like side curtains of a huge performance stage. You *gaze* into the landscape. Two ravens fly into the scene from the edges in a choreographed pattern of swoops and dives perfectly matching the rhythm of your playing. Everything on this natural stage is in harmony with the sounds of your *seeing catcher*.

From your overlook, you are able to *see* across the vast landscape, through the corridor of the canyon… all the way to the far purple mountains. It is a spectacular day. A recent monsoon has washed the entire desert into fresh colors and magnified the vibrant scents of sage, rosemary and lemon grass. You hear the Voice of *Seeing*:

"See your natural senses as super powers; see them as organic tools of perception, able to expand your knowledge of this beautiful place and open the crack between the worlds."

Look out to the far mountains and remember a not-doing that combines Energy Body flight with both open eyed *gazing* and then closed eye *seeing*. This not-doing bridges two worlds using the totality of your visual capabilities. You have performed this not-doing since you first started showing your daughter how to be a perceptual magician. She only needs to stop her internal dialogue and be prepared to journey to those far mountains with her Energy Body. Then, slip through the crack between realities.

You instruct your daughter to *gaze* at a distant section of the mountain range with open eyes. You remember the time from long ago, when your daughter was very young, and she discovered how to look at the tip of her nose to cross her eyes. Learning how to *cross* your eyes is one of the first mastered *super powers* of a young magician apprentice. You instruct her to cross her eyes. You are both *gazing* into the distant mountains, now heavily shadowed with the angle of the setting sun. You both find and agree to *gaze* into the same dark hole in the mountain and to keep your eyes *crossed* but focused on the strange image within the overlapped area.

This nebulous area of overlap, where things are difficult to interpret through your reason is the location of the *crack between the worlds*. Your world of normality is gently moved into the mystery of *seeing*. The "crack" is a fluid watery passage created where the two dark holes from the mountain landscape overlap at the center of your *seeing* using crossed eye *gazing*.

You continue to play your *seeing catcher*. After a short time, you hand the instrument to your daughter. She plays the *seeing catcher* with ease, pacing the speed of her playing from slow to faster, increasing the bellow movement with perfect timing for spirit flight. When it is time for takeoff, she accelerates the rhythm faster and faster. The tone and its cadence are perfect for Energy Body flight. After a short time, both of you express that the *crack between worlds* is open in your crossed eyed *seeing* and your daughter once again instructs with the words:

"Let's Go!"

Your destination has been established with your open-eye *gazing* and now you are ready to fly with your eyes closed. You both close your eyes and fly toward the mountain with your Energy Bodies. This time you are spreading your wings further, secure with each other's flying prowess. You have agreed to approach the crack in the distant mountain that was established with open eye precision.

The mountain gets closer and closer, until the dark crevice expands, and you both are engulfed in a large passageway. All the details of the rock textures appear in focused clarity as you speed deeper into the passage. You intend a shrinking of your Energy Body into a fine line as the passageway narrows. You fly as far as possible before reaching a small waterfall at the very end of the corridor. The scent of water against the granite, and the spray of fine mist envelop your *seeing*. A fog of watery lines encompasses your vision and you suddenly open your eyes.

You are both back at your original place before the flight. You have timed your return by paying close attention to the change in energy that occurred when you moved into the waterfall. The cold shower was an abrupt wake up call. And your daughter stopped playing the *seeing catcher* at that precise moment, bringing you both back at the instant the sound carrier ceased. You talk about the importance of being in two places at the same moment. You compare your *seeing* notes with your daughter to validate details of the experience. Although she was projecting her Energy Body and sensing the far mountain, it was of equal importance that she was aware of her sitting physical body.

The awareness of being split was one of the most important lessons in the Toltec teachings. The Toltec considered the focus only on the Energy Body as an indulgence that would make the mystery of flight into another frozen assemblage point. On the other hand, if the apprentice remained cognizant of being in two places at once, then they kept their assemblage point fluid, and that fluidity was the essence of success.

Your daughter is growing up. Soon your walks together will become less frequent as she ventures into the world on her own. Today she has moved another step closer to a complete perceptual magician. The use of the mystery of the Energy Body is a super power passed from benefactor to apprentice, from father to daughter since the beginning of the Quetzalcoatl Lineage.

You remember when she went for her first walk with you. You told her a magical story about a little girl who had a friend who was a rock. And then she found a stone that she brought home to place by her bedside. A small crack in the stone was the location of the stone's mouth; the place where the stone would talk in confidence to her. That crack in the stone was an opening to her imaginative super power. The ability to hear a rock communicate was the first moment she crossed the parallel lines from one world to the next. You remind her of your memory and end this walk by saying to her:

"I have always seen, from the very beginning of your life… that you were a super hero."

Not-Doing: Hearing as Seeing

Sit in your Power Spot and close your eyes. Begin an auditory search, a scanning with your *ears* for a particular sound within the landscape that resonates with your *wish* to travel with your Energy Body. Once you discover this dominant sound, focus all your attention on its nuances, its direction, and its qualities with your eyes closed.

Keep your eyes closed and imagine that you are traveling on the current of that sound. Once your focus has been clarified, imagine your Energy Body escaping from your physical frame. Some perceptual magicians *see* the Energy Body leave from the top of the head, while others *see* it leave from the forehead or the solar plexus at the center of the physical body. The importance of this not-doing is realizing that you can sense the outside world without "looking" and give flight to your totality; that you can *see* with the eyes of your other body… the Energy Body.

See details of the landscape move beneath you. Keeping your eyes closed does not mean you cannot turn your head in all directions for a better view. It is fine and even helpful to move your head around when using your closed eyes, just as you do when viewing with your open eyes. Keep relaxed and remember that this is a not-doing. You are *seeing* with your closed eyes. Although this is something that is not normally considered in dreaming, once you become aware that you are always using your sight in this second attention, a piece of the totality of your capabilities should be easily understood. Think of the Energy Body, as the other part of yourself that is used some nights when you have a vivid lucid dream. Both bodies have the same senses, but the mobility of the Energy Body is far more advanced.

Continue to *listen* with eyes closed when the particular sound you are using as a guide becomes integrated with many other sounds in the expanding audio landscape. Visualize that all the sounds are a unified accompaniment matched to the speed of your traveling Energy Body. *See* the landscape all around and remember as many details as possible. Intend that one of your *super powers* is to feel your physical body while sitting with closed eyes in one location, and at the same time your Energy Body is flying over the landscape. Accepting that you can be in *two* distinct places at the same time is one of the ultimate *super powers*.

This *not-doing* can also be practiced in the city. It is not necessary to be in your Personal Power Spot, but only in a protected location. Your Energy Body can just as easily move over a cityscape when you *wish* to follow a particular sound in the city. Pick a dominant sound around you that is able to carry your attention out into the far corners of the city. An easy not-doing is to follow the sound of a passenger plane. *See* how long you can follow the sound of the large plane flying overhead until it is no longer available. During the time you were being guided by this sound, your normal internal dialogue should be momentarily vanquished and replaced with silent knowledge. Silent knowledge is another name for the Voice of *Seeing*.

Not-Doing: Crossed Eye Seeing

Sit in your Personal Power Spot in wild nature so that you have a view of the distant horizon. It is best that the time of day emphasizes the shadows in the landscape, during early morning or at twilight at the end of the day. Find a spot in the distance, somewhere with deep shadows crossing the terrain; a place where there is a contrast between dark and light areas. The area where you concentrate could be a dark cave, or a more pronounced shadow from a large boulder that casts a long shadow over the landscape.

Focus on this area of contrasting dark and light in the distance. *Cross your eyes* far enough apart to make *two* images of the same contrasted area. Continue to move the images far enough apart with your *crossed eye* vision so that they *overlap* at the center. Concentrate on this overlapping central area. Consider this *overlap* as a slit, as a crack between two worlds, and that this area can be the passage for *dreaming*.

Imagine your Energy Body lifting out of your physical body and then with your "will" push the Energy Body out and forward from the center of your physical body. You have established your target area with your open crossed eyes, and can now close your eyes and use the *seeing* capacity of your Energy Body. Fly toward your destination. When you reach the crack between the worlds, it is important not to indulge in a prolonged exploration of the area. The Toltec phrase "the crack between the worlds" is a way of naming a mysterious region that cannot be defined in the world of reason but nonetheless forms a visual passageway between separate realities.

As children we enjoy crossed eye viewing, but for the Toltec it was much more then an amusing phenomenon. To the Toltec, viewing with the two eyes overlapping, created a mysterious vision that they used in order to stop their internal dialogue. Without accepting syntax out of the Known, a perceptual magician will be unable to realize the Nagual. The Toltec name for the Unknown was the Nagual. The Nagual could be used in reference to entering the crack between worlds, as well as for a person who was able to bring others into that parallel world. If you have chosen and intended to reach a consensus in the Unknown, then a total restructuring of the Known or tonal naming must be considered.

Without this movement of your assemblage point, the entire path of the perceptual magician will look as mere folly. Only through a disciplined shifting of their vision, will a person ever come close to reaching the inconceivable worlds of the perceptual magician. Only through accepting a new syntax of thought, will an intended person ever reach the Nagual.

This not-doing should be practiced slowly without indulgence. Over time the perceptual magician uses their Energy Body for very specific tasks. They conserve their energy and maintain a not-doing intention to find an answer in the distant mountains to a very particular problem. Flights of fancy are not available to the accomplished perceptual magician. The idea of the "crack between the worlds" was also a Toltec reference to understanding the many worlds that exist, like the multiple layers of an onion skin, around our own Earth density; numerous worlds separated by a foggy region that hid one world from the other worlds.

The practice of *gazing* was a not-doing that opened the idea of *seeing* the fluid membrane that divided worlds. The limbo area between the worlds was the watery passage forming a boundary between worlds. The *crossed eye* not-doing was a reminder of the dimensionality of our *seeing* due to the physical location and separation of our two eyes. The placement and distance between our eyes was in itself a physical representation of constantly uniting two entirely different perspectives. The Toltec surmised that just as our eyes give two subtle, different and distinct views of the outside world, so also there is a separation between what we view, and our own physicality in a particular reality.

The art of *distance gazing* consists of many different methods developed over time by the Toltec. Some Toltec suggested that one should always scan the distant area rapidly when *gazing*, while others preferred a *stilled* eye focus for greater intensity. Some magicians preferred when *gazing* to use their eyelashes as filters interacting with sunlight. They would squint at the bright sun until an iridescent web of rainbows laced their field of vision. These perceptual magicians insisted that *gazing* through a curtain of eyelashes prevented unnecessary distractions from entering their field of vision. In addition, the rainbow web was *seen* as a magical alignment of human abilities with the life-giving rays of the sun.

Particular care was taken with each apprentice regarding any *seeing* technique. Each not-doing was treated with respect and the classroom set up with proper intention: placing the *gazer* with some form of back support, special cushions, appropriate silence, or providing a *seeing catcher* sound. Since these were not-doings, indulgence was considered detrimental to the process. For the apprentice it was a fine line between *gazing* and possibly going to sleep in the process of sitting for long periods.

The teacher would sometimes use a "water rattle" *seeing catcher* behind the head of the apprentice to hold their attention while they were dreaming with closed eyes. The water rattle was a hand-held container with a tight lid, filled half way with water. It was shaken behind and around the head of the student to keep them focused, by following the sound of the rattle. The sound of sloshing water moving from the left ear to the right ear, and then above and behind their head was a method of keeping an apprentice alert for long periods of group *gazing*. It was also a magnificent *seeing catcher* using a completely organic sound, the natural sound of water.

Both open eye exterior *gazing* and closed eye interior *gazing* were affectively augmented with the sound of shaken water. The internal dialogue was erased as the apprentice was forced to follow the auditory progression of the water around their head. The water sound would be near and then far, minimal and then erratic, depending on the creativity of the player. The person playing the water rattle held the intention of water, and moved from fast crashing waves to the slow gentle sound of a brook with their imitation of water sounds. By following the various sounds of water, the apprentice began to feel and *see* images of water with closed eyes. The water rattle was one of the first *seeing catchers* to be utilized by the Toltec.

Not-Doing: The Seeing Catcher

Find a *seeing catcher* in a store or create a personal *seeing catcher*. In either case, omens and spirit directives from the Voice of *Seeing* must be followed when procuring or making a *seeing catcher*. The success of your enterprise is when you use closed-eye *seeing* while you are playing your instrument, and you are validated by the Voice of *Seeing* with regard to your selection.

The ancient Toltec used a drum, an ocarina or rattles as *seeing catchers*. A *seeing catcher* is a "portable" sound device or small instrument that you can easily take with you to your Larger Power Spot. It must produce sounds that are not dependent upon musical nomenclature or structure. The *seeing catcher* is primarily a "drone" instrument that helps to shut off the internal dialogue. The *seeing catcher* encourages *seeing;* meaning that the listener is not counting out beats or waiting for a Known melody. A *seeing catcher* helps you move on the stream of sound with your Energy Body.

Finding a personal *seeing catcher* should be a not-doing. I found my device when I wasn't looking in a children's store. When I was randomly picking up children's instruments and began squeezing the small accordion, I instantly was moved into another space and time. I knew that this was not just an instrument… it was a sound tool for dreaming.

Play your *seeing catcher* with *eyes open*. Watch the landscape of your Power Spot. Consider your playing as the overture to a great opera that will be performed on this landscape stage. You are accompanying the dance of the birds and the leaping deer. *See* the Play of animals, the trees, and the insects all performing within your bubble of sound making.

Now play your *seeing catcher* with *closed eyes*. The manner in which you play your *seeing catcher* should not involve thinking. A *seeing catcher* should feel like it is playing itself. For example, I no longer have to think about squeezing the bellows of my small accordion together because my hands are on autopilot. Your sitting body should be playing an instrument that plays itself without the mind interfering with a song or any established melody. You will use the power of sound to take you into *dreaming*.

Dreaming is an essential ingredient for becoming a total human being. The Toltec believed that as much as bodily exercise was necessary for the physical body, so also was the importance of exercise for the Energy Body. The Energy Body was *seen* as a pure perceptual unit capable of going through walls and out to the universe at large in dreaming. When the Energy Body was complete and strong, they surmised that information about life could be retrieved and used from the dreamtime just as it is collected in waking time.

During a lucid dream, the Energy Body was engaged with maximum *seeing*, since the eyes were the primary sense collecting the dreaming data. The Toltec could *see* that the Energy Body had large dark eyes, vast pools of absorbent intention. They determined that the "will" originated from the center of the dreamer and was the motivation necessary to leave the physical body while the eyes were the collection center. They could *see* that the eyes remained after the death of the body, and were developed into the main component for navigation and interaction with worlds after death. Just as the eyes were the main sense used to collect information on the Earth, so they would evolve further after death into maximum *seeing* with a greater range of sensing color, shape, and precise focus.

In order to assist apprentices in exercising their Energy Body and their Energy Body eyes, they used the sounds of a *seeing catcher*. They could *see* the Energy Body leave the apprentice and they traveled side by side with their own Energy Body to instruct the apprentice in those distant locations. Because the flight of the Energy Body was instantaneous, they considered this practice a natural human ability that had been relegated to obscurity due to societal constraints. It was the obligation of the Toltec Lineage to bring the knowledge of the control of the Energy Body into the light of day. The importance of bringing back visual information, collected in the landscape, was proof of successfully exercising their Energy Body eyes.

The development of sound instruments throughout the globe that can produce the "drone" sound, are the remnants of the ancient perceptual magicians' explorations in using "sound" as a *super power* fuel for dreaming. Combining sound and vision was one of the main components of the teachings of perceptual magicians worldwide. True civilizations contain expression of sound as a vehicle for transcendence.

THE POWERS: SUPER HEROES

V

The Gifts: Wishing Stone

When a small heart shaped stone was discovered in wild nature through a *wish,* and resonated when held to the heart or the forehead of a Toltec apprentice, it became a type of key that unlocked the door into the wonderland of personalized *seeing*. This stone became the apprentice's life-stone, and they were asked to *gaze* into the surface of the rock to find a path for their life in the random veining and coloration of the rock. When an apprentice meditated on this power stone, they began to *see* answers to their life questions. Perhaps a dark spec was *seen* as a doorway that led to a lighter section in the rock patterns representing an open field of freedom. This stone was a gift from the spirit that resonated with the apprentice's *wishes* and desires for awareness.

To find a power gift was of great importance as a visual aid in the conquest of one's fears, when the path of magic was too difficult, or to validate a particular moment in the walk. It was always *seen* as an omen of truth. This stone had been formed by intense heat coalesced into a heart shaped artifact that would stand the test of time, and be with the warrior till the end of their days.

The mastery of "will" is the hallmark of a warrior of perception. The apprentice was taught to have the "will" of a stone, able to continue with the hardships of learning and maneuver through the gates of fear, clarity, and power. "Will" is the magical force of motivation fueling the path of the perceptual magician. Without the intense passion of the "will" to succeed, there can be no path of Toltec knowledge. The path of the Toltec magician is called the path of unbending intent... the intent to use the greatest of your super powers... your "will" to continue, no matter the hardships, in the quest for awareness. An apprentice held the personal stone as they walked great distances to inspire fortitude, or *gazed* into its surface for answers during moments of confusion. The "unbending intent" contained in the stone, to survive any obstacles and continue to inspire an apprentice with its presence, represented a gift of love from the Earth itself.

It was important for the Toltec to "mark" areas that had significance for their *seeing,* with stones that would be *seen* by other magicians in the future. They used heavy stones because, of all the natural materials on Earth, stones are products from the birth of the planet, and had the longest longevity and durability. They considered stones as the oldest library on Earth. The perceptual magicians arranged stones in circles or patterns that aesthetically complimented the natural landscape. They found out that many people, distracted by their internal dialogue, would walk right past a magical circle of placed stones in nature. Walking with an intention to *see,* allowed those who had *wished* to *see,* an open opportunity unavailable to others on the exact same path. The Toltec intended that only those who were talented at *seeing* the markers of other perceptual magicians, could *see* the significant rock markers that blended-in perfectly with the environment.

The Toltec were the designers and artists of MesoAmerica. It was their influence that shaped the ideas of many of the cultures in that area including the building of Teotihuacan, one of the largest cities in the region. The Toltec had moved thousands of stones in order to build their pyramids, which resembled the sacred mountains on the horizon. The Quetzalcoatl Lineages of the Toltec were the Nagual teachers of perception. They used the giant pyramids of Teotihuacan as a classroom to teach the intricacies of *seeing*. Sitting on the stone steps of the high temples, or walking through the wide streets, was the moment when the apprentice had been invited to complete their studies with the masters of perception, in a classroom designed to demonstrate the power of Toltec "will". These large structures of stone were built with the impeccable intention of guiding future generations of magicians.

When the apprentice was ready to *see* the map of the origin of the Toltec Lineage, they were led to an underground corridor beneath the pyramid of Quetzalcoatl in Teotihuacan. The underground passage was accessed through a stone door positioned at the lowest tier of the exterior pyramid. Once inside, the apprentice was walking into the underworld, into the stone heart of the planet, and everything in the tunnel was arranged to enhance the apprentice's *seeing*; pyrite balls lined the passage, glowing from the flame of torches; a stream of liquid mercury marked the border between the middle world and this lower world.

In the central chamber right beneath the pyramid of Quetzalcoatl stood the Four Nagual Fathers, the ones who had created the human race at the dawn of time. They were sculpted in greenstone and stood in the four compass directions facing upward. They were staring at a metaphoric central column of blue energy that linked the lower world with the various worlds above the pyramid: the Sand World, the Water World and the Air World. Each Father had particular necklaces that hung from their shaman shape forms, but only one of the statues had arms and hands and a bag on his back that held the various tools of perception, including the flying disks used to access these other worlds. Arms and hands were not necessary to manifest the human species, only the intention from the eyes. Each Father had reflective mica eyes embedded in their stone surfaces to represent the evolved "*seeing*" necessary to create the human race.

Placed before the four statues were sculpted stone carvings that depicted the "three seed stones" of awareness that activated the mold of man; a cutout form that held the framed energy of a human. The apprentice was shown the stone representation of the specialized tool that was used to crush the seeds, allowing the blue light of life to be released into the mold, stirring the energy that would eventually manifest the human species.

When the apprentice was led into this chamber they were being shown the answers to the origin of human life on this planet, and they were ready to become Nagual leaders and eventual world builders themselves. The Toltec believed that a Nagual would eventually leave for a mission that involved using all the *seeing* information they had stored while on Earth. It would be their ultimate task to "seed" other worlds into awareness. The Nagual would remember all their *seeing* lessons of air, water, and fire and the mysteries of colors and shapes, their catalogue of plants and animals, the density and varieties of stone and the various minerals on this planet, and even the understanding of love, in order to begin new earths in distant galaxies.

....

Much time has passed since the first walk in the city with your daughter. You are the father of three children now: one grown daughter and two younger sons. Your walks involve an extended bubble of awareness that must *see* the needs of all three personalities. You begin this walk in the city, where you began when your daughter was very young, in the skyscraper that you used for the first *seeing* lessons about water moving its facade. You step into the exterior elevator with your three children.

It is a glass walled elevator that carries you up the exterior side of the skyscraper through the area of the rippling water; the place you and your daughter activated long ago when she was only eight years old, with the "moving river" magical not-doing. Your daughter, who is sixteen now, also remembers and sways her body in tribute to the memory, as if she is crossing the rough current of the river.

You are also here with your two sons. You hold your younger son's hand tightly and instruct him to just close his eyes if he is afraid of heights, as the glass elevator moves higher up the building side. But he is not afraid and even moves to the edge of the elevator floor for a better look down at the shrinking perspective. And your son *wishes* confidently:

"I am a flying eagle."

When the elevator stops, you get off and walk into an expansive roof garden, open to the evening sky. It will momentarily become your family's Personal Power Spot at the top of this manmade mountain. A section of central lawn is dotted with carefully placed boulders that create a natural looking landscape. Your oldest son *sees* something and walks over to one of the large stones and discovers that someone has rolled up a small piece of paper and placed it carefully within a narrow crevice in the boulder. He pulls the paper out of its hole and unrolls the paper. There is nothing on the sheet and so your son suggests that you write a message on the paper for someone to find in the future. When you all agree on the best sentence to write, you take out your pen and carefully write these words:

"All my wishes have been granted."

Your son carefully rolls up the message and places it back in the stone. You all look at one another and feel that you have written the greatest message for someone in the future to discover. Your daughter believes that the *wishes* we make for ourselves are the *wishes* that we make for everyone. The Voice of *Seeing* says:

"Omens may appear all around you, but their meaning doesn't always appear in that instant.

The gift of deciphering was given to you so that your journey on this planet is extended. The gift of forgetting was given to you so that your journey is filled with surprises. Only when the long walk in the Wheel of Time is complete, can the story of your life be seen as a wondrous Play, perfectly scripted with a beginning, a middle and an end."

You *see* that everything you have taught to your children is returning in a great cycle of creative demonstrations. They are skilled enough to be your guide, and you are wise enough to become their apprentice. Your children have moved from apprentices into neophyte teachers. You trust them to enhance your own *seeing* into realms that are only possible by becoming a beginner yourself. And the Voice of *Seeing* continues:

"When the teacher becomes the student then the path of awareness has widened. Your children are also your teachers; they have shifted you into seeing hidden aspects of yourself. When the perceptual magician understands the super power of sharing and the power gift of receiving, then the path is paved with true joy."

Your son finds a straight branch lying at the base of the boulder. Your daughter and you watch as your two sons climb to the top of the boulder and stick the branch into a crevice at the top. Your older son proclaims that everyone in this office building will have all of their *wishes* be granted in the future.

You hold hands together in a circle and then lift your arms skyward. All four of you fall down onto the grass and lay facing upward to *see* the night sky of twinkling stars. In the map of the Toltec, the night sky was *seen* as the gigantic dark eye of an eagle watching over all creation on the planet. The black eye of the Eagle is *seeing* all of you. Remembering that all creation is linked together by perception, and that in the Wheel of Time there will be a return to this moment in your awareness, it is time to mark the occasion with a not-doing. You instruct your children to wave their open palms in front of their eyes in a rapid crisscrossing of fingers and then to drop them suddenly to reveal the dark sky. Everyone *sees* a grid of phantom crossing lines floating in the night sky. Then the Voice of *Seeing* explains:

"When seeing is shared with others, you have bestowed the gift of creativity. The luminous lines forming a grid across the sky are made from the same energetic emanations that crisscross through all life on this planet. These fibers of light are the true connectors uniting all things."

You reflect on the privilege of *seeing* all the wonders available on this small blue planet floating in the great dark sea of infinity. The Eagle watches over the Earth and all its inhabitants, just as you watch out for your three children. Both of you *see* the necessity of bestowing awareness on all those you love. It is not easy to create worlds of happiness, but the effort and the challenge is worthy of a warrior and a Toltec father.

THE GIFTS: WISHING STONE

You wonder about your children's future. You wonder if they will remember these times together... times that are so important to you. And then you *see* that these magical times will be the ones they will remember and cherish for all time.

The branch, which had been precariously balanced, falls over, rolling down the rock face, and lands exactly at the base of the boulder where your son had first discovered it. Your oldest son explains that a branch with such unbending "will" to return to its place of origin can't be considered as a regular piece of wood. He proclaims that the branch is now a magic wand. It holds power. The power to move at "will". Your son waves the wand through the air and then places it in his belt. The branch has transformed into his personal power gift. And you hear the Voice of *Seeing*:

"Every object in nature has power. The things you 'see' in nature that pertain to your wish are the objects that represent gifts of power for your journey. The 'will' to survive and flourish is the motivational power behind all life on this planet and in the greater universe."

Your two sons are already climbing back up to the top of the boulder. This time they plant the wooden wand into the slit and pack the lower edges with smaller stones to keep it steady. Your oldest son names each of the smaller stones as he carefully places each one of them around the base. He names one stone "care" and another stone "balance". You are *seeing* yourself in his actions. He is giving back the gift of his creativity to his father. And the Voice of *Seeing* says:

"A true power gift is something of personal value that appears outside of the Known. It is something that can reappear or disappear when the power has left, or when it is no longer needed. But the naming of the gift is a way of always remembering its power."

You go back down from the roof garden inside the glass elevator. You *see* that your children are happy to contribute to the lineage of perceptual magicians. You *see* your youngest son holds out his arms, being an eagle again, and he stares straight down as the elevator descends along the side of the skyscraper. He folds his arms back to his sides when the elevator touches Earth. And then the Voice of *Seeing* says:

"Your children are growing toward their destiny and being able to see them fly free with awareness is one of the rewards of a Toltec father."

Not-Doing: Finding a Power Object

Go to wild nature or to a city park that has become your Larger Power Spot. Make your intention "out loud" to find a *power gift* of a life-stone. A *power gift* is any small object that you can hold in your hand that carries the weight of spirit intention. It is usually a beautiful manifestation created by nature: a smooth stick or palm sized stone, or something originally from animals, such as a feather or a deer antler. A power gift can be from the ocean, like a shell or a piece of water-sculpted driftwood. The range of power gifts related to the Earth is endless, but they always pertain to the particular quest you are walking. The spirit has given you this *power gift*, and you honor its discovery.

I was walking in the mountains of northern Mexico when I came across a sun bleached coyote skull in the underbrush. It is very rare to discover a complete skull, so I spent time dusting off the caked sand and then scraped away any residual fur. When I got home that day, I received a phone call informing me that my mother had just died. I painted the coyote skull bright red and placed it on my altar. The skull was a power gift that represented the transition between life and death. The discovery of the skull was tied to the death of my mother, and the process of decorating it was a way to ritualize the moment forever in my Wheel of Time.

This day your intention is to find a natural life-stone. Begin walking through your Larger Power Spot. You are scanning the ground for a stone that stands out from all the others. The Earth holds many natural treasures. Small stones have been given as repositories of truth since the molten creation of this planet. The Toltec considered stones to be the oldest library of information on the planet. Your stone has been waiting for you to find it since the beginning of time. It has been waiting for you to come along on the path and read its etched messages.

Finding a stone that is a *power gift* in the landscape is a tangible reminder of that particular day. It will always bring back a memory in the Wheel of Time. Some stones will remain with you for your entire life. These are called life-stones. You can *gaze* within their surface and *see* many trails and detours meandering along the surface. You will be able to read personal messages within the textures of the stone as a map of your life.

As you walk, reflect on your life being revealed in the weight and texture of the stone. Ask yourself if this *power gift* will give you the necessary information to continue on your life path toward awareness. This stone will stand out from all others on the path. Pick up your life-stone and hold it to your heart. Look deeply into its surface and feel its special shape, *see* its unique color, hold it up to the sun to collect the energy. Place it against your skin and let the heat move into your body. Observe all the details and *see* the abstract marks on the surface of the rock revealing a picture narrative of the twists and turns in your life.

It was always important for me to find a *power gift* stone on my way into the large landscape of my Power Spot. When I located the *power gift* in the form of a small stone, I would ceremoniously place it in the same large boulder at the entrance to the landscape. I considered the *power gift* as the key that would unlock the upcoming mystery. By placing the small stone with intention upon the large boulder, I was announcing my respect for the wild Power Spot and ceremoniously opening the door to my house of wonders.

Long ago, I was walking along a mesa in Chaco Canyon New Mexico. I had collected three stones that represented my three children. When I arrived at the edge of the mesa, overlooking the entire panorama of the desert landscape, I was overcome with the presence of my children in those simple stones. Each one of the stones had a special color and shape. I placed all three in a smooth bowl shaped area of a large boulder. That day, I *wished* for their safe journey in life and then made sure that the wind couldn't reach the stones and move them out of the bowl. I wanted them safe for all eternity. I knew that one day I would return to that exact spot. I would *see* the three small stones still secure within the bowl. I would *see* myself as an old man then, and I would *see* that I had protected and loved my children till the very end.

Not-Doing: Seeing the Luminous Grid at Night

Go to your Larger Power Spot at night. Lay on your back in order to *gaze* into the night sky. Take your open hands with fingers parted, and wave them back and forth rapidly, crisscrossing over one another in front of your open eyes, as close as possible to your face for a minute. You are *gazing* beyond the open changing holes formed by your fingers into the void of the night sky on the other side. Then quickly take your hands away to *see* the entire night sky; *still* your eyes within the panorama of the night sky. *See* the phantom fibers of energy that remain; the subtle lines transverse the dark sky, forming a huge grid of crisscrossing lines and small shadow squares.

This not-doing was an important introduction to exercises that the Toltec called the Magical Passes. The practice of hand waving comes from the first set of Magical Passes channeled by the original Toltec. These simple hand gestures opened the door to the second attention, the place of not-doing attention. These passes helped the apprentice *see* for the first time the energetic makeup of the planet. The revelation of after-images floating in the air and then just as suddenly dissolving, activated the desire for further discoveries involved with *seeing* in the apprentice. By waving their open fingers in front of their eyes, they activated a *seeing* of the energetic field around the Earth. These soft lines were the pulsating evidence that their planet was wrapped with an energetic cocoon of protective emanations. The Toltec believed that they were *seeing* what they named, the emanations of the Eagle. This giant guardian at the threshold after death was responsible for recycling life energy. The grid lines around the Earth were also an energy field that originated with the Eagle. The Eagle's emanations formed fibers of light that protected the interior surface of the Earth with this luminous shell.

As a young man I was fascinated with the mechanics of *seeing*. I explored scientific journals and executed experiments concerning the human eye. Some days I could be found walking the streets blindfolded to experience my world without the gift of sight. One of the first incidences of perceptual magic occurred before I was able to *see* the emanations of the Eagle, when I waved my open hand across my face in a slow movement while *gazing* into the sky during the day. I kept my *stilled* eyes focused into the distance and allowed my hand movement to transverse the foreground without distracting my attention.

This was simply a graceful out-of-focus movement, going on, while I was continuing to *gaze* into the bright blue sky. I didn't follow my hand with my eyes. The magic happened when a series of after-images of my hand were detected trailing behind the solid hand. The slow style and grace of my movement enhanced the success of *seeing* these phantom hands. My reason explained this as just an effect of persistence of vision, but the Voice of *Seeing* said this was a moment of choice between the Known and the Unknown. If I picked the Unknown, then I was on the road toward becoming a perceptual magician.

One of the first not-doing passes taught to me by my benefactor came from the original series by the Toltec of antiquity. When watching a warrior remove an arrow from the sheath on their back, they observed an optical oddity. The act of pulling the arrow caused an after-image of the line of the shaft across the bright blue sky. Later, they devised a not-doing by having the apprentice slowly act out the action of removing an arrow across their field of vision, while *gazing* with *stilled* eyes into the firmament at the passing arrow.

The apprentice learned to *see* the trailing of ghost lines from the shaft of the arrow with their peripheral vision. The apprentice was not following the movement of the arrow, but was witnessing the phantom images that happened after the solid arrow had left the previous space. The apprentice was *seeing* how energy flowed in the universe, and also had a greater understanding of the emanations of the Eagle because the fibers of the Eagle remain invisible to the looking eye, but visible to the *seeing* eye.

Not-Doing: Understanding the Eagle

Go to your Personal Power Spot at night. *See* the dark night sky as something out of your normal frame of reference. Instead of a Known description, *see* the dark sky as an ancient Toltec involved with the mysterious world of the Unknown would. Silence your internal dialogue.

Imagine that the whole night sky is the gigantic black "eye" of an immense Eagle. All the stars are but drops of reflected water floating in the aqueous humor of this eye. The bright spot of the moon is a reflection off the central pupil. The Eagle was the Toltec guardian of the universe; the power that everyone would approach at the end of their life with their Energy Body. This power was named the Eagle because in the natural world, the bird able to *see* the furthest was the flying eagle.

This gigantic Eagle stood at the threshold between the worlds of the living and the dead. Those who had developed their "will" within their Energy Body and had practiced dreaming throughout their lives, passed into the other side through its gigantic all-*seeing* eye, while all others who died were recycled into unconscious energy within its talons. Once this information was revealed to the Toltec, it became imperative to them to perfect their "will" and exercise an aware Energy Body. As perceptual magicians they wanted their explorations of the universe to last as long as possible even after physical death.

With teams of perceptual magicians, the Toltec could *see* that all the fibers of creation had their origin at the Eagle. They could *see* that everything was connected with fibers that emanated from this power and animated life on this planet. The Eagle was the caretaker of all the energy that flowed in the Earth, and everything was returned to this power at the time of death.

The Toltec dedicated their art of *seeing* to understanding the luminous fibers that encased the sky. They could *see* that this grid of luminous fibers was a protective field of energy that allowed all living things on Earth to thrive in a safe place. *Seeing* the stability of the fibers was a testament to the *wish* of the Eagle that this world was protected and aligned with the emanations of awareness.

Not- Doing: Activating "Will"

There are many obstacles placed on the path to awareness; ways of challenging the apprentice to activate and begin strengthening their own "will". When my youngest son was learning the magical task of swimming, I used my *seeing* to develop creative Play around this activity and make it into a not-doing. Learning to swim is connected to anxieties about death. A young apprentice is facing a situation that has death at its core. They are courageously approaching something much deeper then just learning how to flap their arms in water. As a Toltec father, I made learning how to swim into a *not-doing* connected to *seeing* and therefore to awareness.

"Will" is a silent super power. It is reached at moments of intensity, when everything seems to point toward failure. "Will" comes from the center of the body and enables a person to perform magical feats that hitherto were unavailable to them. "Will" is connected to *not-doing*, because it comes from a different attention, other than the place of reason.

Take your young child to a swimming pool in the city, or to a lake in nature. Arrange a Play that involves "saving" you, the father teacher from a whirlpool that has you spinning in the deeper water. You have already practiced strokes with your young child, and they know how to swim, but this not-doing is to activate and reinforce their "will." By instilling the desire to move beyond just learning strokes, into the realm of confidence with no internal dialogue, the "will" takes the lead over from reason.

Go to the middle of the pool, in deep water and pretend that you are the King of the Water World, who happens to be trapped in a whirlpool from which he cannot escape without your child's help. Shout out to your child apprentice to come and rescue you from the whirlpool because they are the bravest Knight in the realm. This dramatic Play will help the child lose their old frame of reference, to act spontaneously and come to your aid. In the process of saving you, they have stopped their internal dialogue and become the persona of a "master swimmer" coming to the rescue.

In this process of losing themselves to benefit another, they are activating their "will".

Around the bright fires at night, in the caves of Power Spots and on the steps of the pyramids, the Toltec told tales of power involving the feats of other magicians. These fantastic stories were lessons about the hardships and glorious victories of other warriors on the path to awareness. Their storytelling was a teaching device that had always originated from the Voice of *Seeing*. They were performing a not-doing that maintained a flexible, creative presentation of the material.

As a Toltec father I used a nightly bedtime story that involved characters that continued their adventures each night in fantastic worlds. It was always my intention that the characters in the tales resemble my own children. In this way they understood that the stories were about their personal adventures in various worlds, although I never titled the characters with their real names. Each story involved a little girl or a little boy going into another world to interact with the landscape of Allies and opponents. It was an adventure story with continuing chapters each night, where the characters faced perils filled with humor and learning.

I was guiding them into a dream landscape at night, just as I had guided them on the road to discovery during the day. I was pointing them in a particular direction that would enhance the chances of them retrieving information, while they went to sleep with their Energy Bodies, that would help in solving the mysteries of growing into adulthood.

My stories were a not-doing. Just as walking the trail in the day was an exercise with surprises around the next turn, so also these stories were the Unknown. The only planning was the use of the same characters representing my children. Because this method of storytelling was a not-doing, I had to be spontaneously creative with all the details and the evolving plotline. I had no idea what was around the next bend on the dream road.

It was helpful to add lighting effects to the production for the excitement of interesting shadows and lights occurring in the children's normal bedroom environment, so I often used a flashlight. The use of surrounding toy props with a spotlight was a flexible way to give drama without a *doing* of thought. I would beam the flashlight on various parts of the room or cast the light on some of their toys, contributing excitement to the developing story. I found that my children were always delighted with the creative magic of lighting, and as they grew older they became active participants in the drama by taking a turn as director. This creative Play was an extension of a dream brought into momentary light that would continue when they closed their eyes and went to sleep each night.

VI
The Allies: Imaginary Friends

Finding an Ally was of vast importance to the Toltec. An Ally was a presence that was not connected with the human perspective. It could be an animal, an insect, or an imaginary being outside of the Known. Without counsel available from beyond the "human" perspective, the perceptual magicians had to rely on their own *seeing* or other people, without any validation outside the human frame of reference. And the Toltec knew that the human perspective could easily be distorted with internal dialogue.

Seeing that plants kept their bodies and spirits alive and vital, they deducted that all life had an internal force; a force that motivated the "will" in all things to survive and function, and in the best sense, support the rest of the Earth family. The plants were *seen* as friends who gave life and wisdom and so they called them *Allies*. They reasoned that animals gave them life and production of food products, and were also considered trusted companions, working side by side with the tribe. An apprentice could find wisdom in the singing insect or in an omen of the sudden appearance of a flying hawk. Life on this planet was assisting their walk with energy giving affirmations, if only they could *see* these animals or plants as Allies.

As their studies progressed in dreaming with their Energy Bodies, they met Allies who had no temporal form, but nevertheless assisted them with counsel and trustworthy advice in the dream. These inorganic entities were available friends that could be consulted while they walked in the day as well as in their nightly dreaming. The Toltec apprentice was encouraged to find at least one of these dream Allies and accept their non-human knowledge for a greater range of *seeing*.

In return, the apprentice would take the Ally into their world and allow the inorganic Ally to *see* through the eyes of the human magician. This exchange of knowledge showed the Toltec that the inorganic Allies had existed far longer than humans, and that Ally *seeing* was a storehouse of knowledge with unlimited creativity. The inorganic Allies knew how to take on the organic form of any outside animal, bird or insect, or even take on the form of a fleeting shadow. Just as their spirit presence could inhabit anything in the outside world, they were also able to incorporate inside the human apprentice.

To *see* an Ally, the Toltec went to remote locations far removed from the tribe. They fasted for a vision, and called out for help to find an Ally in order to assist them with the rigors of the Known and the dreaming world. When they located an Ally, it may have come in the form of a special deer that sensed that the seeker was energetically aligned, and intending a trusted friend from outside their normal perspective. Then the magical deer spoke mysteries to the apprentice in a language that came from the center of their eyes. It was communication without the interference of words. Perhaps the Ally came as an insect that landed right on their hand and through the buzzing of their wings, told the apprentice secret methods to make their journey in life more fulfilling. In this way, they found comfort and an interpretation of data that was outside of the human mold.

The understanding that all things on Earth were Allies for one another brought the Toltec to the threshold of *seeing* the flow of energy that maintained a healthy planet. They became aware of their own luminous shell; the egg shaped energy cocoon that wrapped around every human. They could *see* the energy field of light around plants and animals and how this light changed in color when constricted with fear or greed.

One day a team of *gazers* noticed that the Earth itself had its own radiant cocoon. They could *see* the Earth's luminous shell in the glow at the meeting edge of Earth and sky, along the top of the distant mountains in the bright daylight, or in the twilight as the sun was setting. This slight border glow at the edges of the horizon was a validation of a gigantic luminous sphere of energy surrounding the Earth itself. They continued to refine their larger Earth perception. The sky was *seen* as an Ally that brought rain for crops, and brought the magical wind that entered into all living things with a life force. They viewed the sky as part of the energetic field of the planet. Teams of *gazers* were able to *see* a fine set of luminous energy fibers encasing and crisscrossing the sky. The grid originated from the emanations of the Eagle. These lines were the energetic crossings that ran through all life; from the distant sky, to the solid Earth below, through the smallest insect, and through the cocoon of man.

The Red Grid was named for its reddish hue in contrast to the bright blue sky. The same grid lines could be *seen* in the dark night sky by using a Magical Pass. But then the grid was *seen* as transparent lines without any particular color. By overlapping the fingers of their two hands and moving them in rapid crossing close to their eyes, the Toltec activated the *seeing* of the Red Grid.

These specialized actions were not-doings that utilized physical movement to shut off internal dialogue and open a greater range of *seeing*. All these magical movements were first developed in Nagual dreaming, and then, these movements were refined in Tonal awareness. Some were performed with open eyes to *see* the changes in the outside world, and some passes were performed with closed eyes to *see* other worlds. The Toltec were the first biological scientists, exploring the micro and macro cosmos with their bodily sensing tools.

It was the Toltec magician that explored the ability of the eyes to change their perception of the outside world. They interpreted the data sent to their eyes as malleable. The Toltec believed that the images coming to their eyes were not fixed, but through eye techniques could be altered drastically; creating fluid movement inside a solid mass, or flattening a steep path into a flat trail. These not-doing techniques were revealed to the larger tribe in order to change the drudgery of daily life into an evolutionary step for humankind. It was the Toltec who were able to *see* the working soul of themselves and their relationship to the larger world, and use *seeing* to enhance the beauty of being creatively alive for everyone around them.

....

After many walks with your children you have now arrived at the Kingdom of the Giants. You stand at the threshold of a Redwood forest in northern California. Your three children are old enough to enter the next level of *seeing*, the realm of the Allies. To step together across this threshold is to enter the awareness of the oldest beings on the planet, and the place of the legends of fairies.

The children cross back and forth over the small rivulets that meander between the huge trunks and form water trails to another sculpted tree cave.

Some of the huge trees have dark interiors, hallowed out by ancient fires, charred wood caves that could be the homes for fairies or trolls. Your children play a game of hide and seek within this Kingdom of the Giants; these stoic entities have been silent caretakers of wild nature for thousands of years. The coast Redwood Trees have *seen* the time of the dinosaurs, the dawning of humans, and have stored enough energy to be the tallest living species on Earth.

This corridor of gigantic wood pillars leads you deeper and deeper into the Kingdom, until your son spots a cave room at the base of one tree, large enough inside to fit everyone. You sit within the cutout base on a soft floor of furry bark with your three children. This tree room is your Personal Power Spot within the Larger Power Spot you have named the Kingdom of the Giants. And in the stillness of your wooden room you hear the Voice of *Seeing*:

"A perceptual magician needs an Ally. An Ally is like an Imaginary Friend if looked at from a Known world perspective, but it is very real to the Seer's perception as it serves the progression of seeing. A magician without an Ally must rely on their own devices with no one to counsel them in complete confidence. A magician strives to find an Ally, an invisible presence that can manifest between different species and can be trusted completely. Remember in your life when you communicated deeply with your pet dog or cat. This non-human friend could be consulted whenever needed, and was a reliable, trustworthy companion."

You feel the comfort of having deep roots beneath the floor of your cave, and the energy of a ceiling that grows all the way to the sky. This wooden cave is a dreaming space that opens portals of the imagination into other worlds: worlds of the leprechaun and the forest elves. You announce to everyone that today's *wish* is to find an Ally. Your children agree with an enthusiastic nod of affirmation. They are eager to explore the esoteric knowledge of the Toltec and appreciate the creativity of not-doings.

You sense that *imaginary friends* and magical creatures are lurking on your gaming field. You remember when your children imitated the actions of animals when they were very young, often picking a favorite creature that would remain their spirit guide for life. Your children announce that they would like to use closed eyes to *see* where these creatures could be hiding. It is time for you to use sound guidance and take your children into closed eye *seeing*.

You stand behind them and rub your flattened palms together producing a subtle sound that reminds them of the wind moving through the branches of these large trees. By changing the cadence and occasionally cupping your palms as they rub together, a melodious echo is produced, that encompasses your protected hideaway. This not-doing technique was one of the first *seeing catchers* because it was based on bodily resonance without outside mechanical instruments. You circle their three heads, from one of their ears to the other, surrounding them with the sound of gentle rolling waves of swishing sound.

You instruct them verbally to *see* the sound from your hands transporting their Energy Bodies to a vast landscape of amber sand dunes and a bright white sky. You are guiding them to the Domes of the Nagual, the place designed by the ancient Toltec in the far constellation of the Belt of Orion. You continue producing the wind sound with the palms of your hands.

They all travel on the waves of sound intention and begin adding their own descriptions of what they are *seeing* within their closed eyes. Your oldest son starts by stating that he *sees* an immense dark dome on the horizon, a gigantic structure that reaches up to the white sky. Then your daughter adds that she is *seeing* someone gilding next to them as they head toward the dome on a path that is made of rainbow light. Your youngest son *sees* that the being has large glowing eyes and no arms or legs. Your oldest son says:

"Our visitor is from another world."

You remember the tales of guardian angels who are trusted invisible friends watching over people, particularly children. My three children continue taking turns describing their internal *seeing* while the angelic being takes everyone into the huge dome. They describe *seeing* other angels drifting over an immense amber field with green boulders scattered throughout. The height of the dome is immense and forms another type of sky. There are pyramid buildings that have fluffy elevators moving up the sides, and a long beach stretches out by a glittering lake, where inverted pyramids are collecting a blue energy that falls from the dome sky. Then you hear the Voice of *Seeing*:

"Every being in the universe has the potential to become a trusted Ally when they guide you with silent knowledge. The Ally will introduce you to inconceivable worlds, normally unavailable to the human perspective. When you connect with another species, a dog or a cat, or a wild bird, then you are dreaming the same dream and becoming Allies with one another. Fear has been removed from the equation, and at that moment, when the two of you have stopped time in peaceful harmony, in that instant, the world of reason is vanquished and you are both Allies for one another."

You stop the rubbing sound of your hands and everyone opens up their eyes. They discuss more details of their visit to the Domes of the Nagual; how the angels took great care in guiding them throughout the wonders of this new place, and about the beauty of the architecture. Then the enthusiasm is replaced with stillness. Everyone contemplates the immensity of being able to travel back and forth between worlds. Their combined dreaming has had a profound effect.

The silence of this deep forest awakens with the sounds of a Raven bird family. Their shiny black bodies slowly descend from the heavens onto the valley floor. Their voices resemble human chatter and everyone begins interpreting their sounds, as if the birds are speaking directly to each one of them. The Raven family has four members; a mirror of your own grouping. Your children *see* that the birds are discussing your presence, just as you are talking about their appearance.

And then it is time to climb to a higher ridge. When you arrive at the top, everyone takes a position overlooking the beautiful Redwood forest. When everyone is ready and saturated with the magnificent view, you instruct your children to close their eyes once again. You have them turn their heads upward so that the brightest light of the sun bathes their closed eyelids. Then you instruct your children to place their two hands between their eyes and the direct sun so that shadows of their fingers cross over their faces. You tell them to begin overlapping the shadows from the fingers of both hands so that a pattern of crossing shadows and light wash over the closed eyelids. You instruct them to move their two open palms rapidly. The crossing shadows of their fingers are making a grid of dark and light across their closed eyelids.

You are showing your children the luminous Red Grid. Perceptual magicians understand that after practicing with eyes closed, you can see the Red Grid in the sky with open eyes. Have them open their eyes. These fibers of light represent the luminous shell encasing the Earth. It is a depiction in the sky during the bright daylight of the energetic matrix around the whole planet.

Your children open their eyes to *see* into the clear blue sky. This time you instruct them not to look at the sun, but into the blue sky. Then they wave their open palms with spread fingers close to their open eyes. When they take their hands away they are *seeing* the red fibers of light transferred from their inner world to the larger world outside. They are *seeing* the larger Red Grid in the sky. You explain the nature of the energetic field around all living organisms. The Earth itself is a sentient being that produces a protective dome of energy. Perceptual magicians call it the luminous body of the planet. Just as humans have a luminous energy field, so also the Earth has a luminous body. The Red Grid is the Toltec way of *seeing* this structure of energy that encases the Earth's body.

It is time to venture back home. A small stream of water runs the entire length of the steep path along the ridge. Your youngest son has fun maneuvering the grade and enjoys sliding backwards in the particularly muddy sections. You instruct him to keep his eyes *stilled* right in front of his moving feet as he walks forward. This concentrated focus will activate a *super power* within him. In an overly theatrical gesture of confident strides he effortlessly walks up the steep muddy trail. He tells you that as long as he keeps his eyes glued to the tips of his shoes, the path is no longer steep. He *sees* the path become nice and flat. Your youngest son says:

"The world is easier when you use *seeing*. It is easier still when you have a trusted Ally in your father."

You stand *seeing* the entire Redwood valley below, spreading all the way to the ocean. This Kingdom has lasted since the dawn of time, and you are *seeing* its "will" to survive. It has survived fires and floods and the encroachment of man. The Voice of *Seeing* reminds you that one final magical not-doing is recommended after a successful day. Toltec magicians of perception have a long tradition of turning toward the Larger Power Spot at the end of a journey to thank everyone. And you pronounce loud enough so that the Redwood forest below can hear:

> "Thank you everyone: the spirit, every tree, every bird, all the clouds, our angel, the light, and my children. We have practiced our *seeing* today. The gift of *seeing* is a gift without limits. To utilize its super power is to fulfill the *wish* of our totality as human beings. Thank you all and everyone for being our Allies.

Not-Doing: Calling an Ally with Closed Eyes

Go to your Larger Power Spot in wild nature. Begin your walk with the intention *wish* to "find an Ally." Proceed to your Personal Power Spot in the large area. Sit down and *close your eyes*. Play your *seeing catcher* if it is available. If not, then listen closely to all the natural sounds in your environment. Perhaps use the not-doing of rubbing your two palms together to create a gentle sound that tends to filter any internal dialogue with its white noise. Rub your flat palms together, varying the speed and pressure to create the sound of the wind or of the ocean waves. Your body is an instrument. The first shaman drum was the bodily chest cavity used as a resonator when rhythmically tapped with the shaman's palms.

Keep your eyes closed and visualize your Energy Body flying into another space, with the creative freedom of your "will". Intend to reach another world. You can *wish* to travel to your vision of heaven that resonates with your spiritual path in order to discover an Ally. Eventually you will be able to *see* a landscape with practical features: a natural terrain or a cityscape or even an off-the-Earth world environment. Sense all the physical attributes of this world: the trees, or rocks, or buildings. *Gaze* into the space as if viewing the vista from the perspective of a traveler who has just arrived from another world. Then move your *seeing* rapidly from object to object. You are collecting data about this new world with your closed eyes.

Energy Body *seeing* has a tendency to remain fixated on one object in the landscape, so continual scanning is advised. Indulgence creates solidification in the flow of energy, in both the Known and the Unknown worlds. Picture your interaction as similar to continuous walking in your waking environment, with a constant panorama of new data coming to your eyes. You will have the feeling of gliding, as opposed to taking steps with your legs and feeling the impact of your feet. There is no friction, only a smooth transition from one place to the next. The only impediment is your Energy Body's density. The feeling of weightiness varies from the lightness of flight in the Air World to great heaviness in the density of the lower worlds.

Watch if any person or animal or abstract energy form enters your energy-generating world. This *being* is the answer to your beginning *wish*. The appearance of an Ally varies with every perceptual magician. The Ally will appear as an animal spirit or the image of a person or even an abstract geometric shape or as an amorphous intense color, but its appearance should give you a feeling of deep connection and peaceful alignment with a real and practical entity.

Ask this Ally out loud to become your companion. *See* with your closed eyes if the Ally responds in some way, verbal or with body language as a somatic answer to your question. The Ally can respond by simply staying within your field of *seeing* or even performing a magical movement that answers your question about alignment, or with a change of bodily color. Make a detailed inventory of the Ally. What features on its body stand out that define its individuality? Are there colors that are part of its makeup that are unusual?

You are making a Map, memorizing the image that will signal the arrival of your new acquaintance in future interactions. Every time you go for a walk or when you are in your dreaming, repeat this procedure, until your Ally also becomes familiar with this meeting place, either in the dream world or in your physical Larger Power Spot. The Ally is simultaneously remembering how you are *seen* to them in order to develop a lasting friendship and a common meeting place.

For the Toltec, finding an Ally was of the greatest importance. In our time, the mystic realm has been relegated to useless meandering of the imagination. Entering this arena means trusting the Voice of *Seeing* that has been silenced for most people since childhood. Your imagination is the threshold of *seeing*. To the ancient Toltec, there were only the mysteries of the world unfolding with open or with closed eyes. They considered dreaming in other worlds as a natural extension of living a full and complete life.

To find an Ally you are returning to your own creative totality. You will have to move out of the "fashion of the times," where generally the imagination and the dreaming worlds are considered minor diversions, and then you can begin to *see* them as the rest of your reality.

To locate an Ally in a dreamt world requires a leap of faith. It is not an easy not-doing, because it entails reshaping your interior life. Do not be discouraged with a slow start, or if feelings from your traditional beliefs attempt to derail your success.

It is useful to understand that this not-doing has a very practical side… the exercising of your imagination. Without personal imagination your engagement with the Known world is a continual reflection of someone else's dream. By working actively in your interior landscape, you are working with unused emanations of your potential.

Not-Doing: Calling an Ally with Open Eyes

As you walk in your Larger Power Spot, pay attention to any animals or insects that cross your path. You have voiced your *wish* to find an Ally, so you are keenly aware of anything that appears that day, especially any animal or insect or bird that enters your field at a particularly auspicious time. When a creature arrives, stop and wait. A rapid response with stillness moves you closer to the realm of wild nature. Everything in the natural world has a specialized rhythm that corresponds with the assemblage point of nature, which is inherently silent knowledge. It could be a special bird that lands to sing you a song. Perhaps a lizard stops on the path to observe your progress. The smallest insect is often the largest reminder of our mutual walk on this Earth. *See* if this arrival answers your question with a gift; something it does with its body or a sound that will represent its sincerity to communicate.

We often talk to birds and imitate their sound patterns with our own whistles as a method of abstract communication. Imitating the sound of our Allies is a gesture for friendship. Any unusual sound-making or strange bodily expression that is sensitive to the environment is also intriguing to the Allies. The animals are *seeing* your energy and they are experts at detecting the human who has stepped out of the Known. When they respond with patterns of sound or bodily movements, we feel aligned with another species. A bird for example will ruffle their feathers or turn away their head to show they have no fear of your presence. A small wild bird will usually keep a few branches between itself and your position as a protective shield. This maneuver insures that both parties can have an extended viewing time together while being separated with a protective wall of twigs.

If both you and the animal resonate with this exchange then you have bonded with an Ally. Every time you walk in this Power Spot and your Ally shows itself on a rock or in the air or when it crosses your path, it is a welcoming gesture from that entire species. You have made an Ally and their extended family has been told of your friendship with one of its members.

Walking in the high desert I was often met by a California Scrub Jay. This bright blue-feathered medium size bird was interested in my arrival into its desert world. Each time I entered into this area, it showed up to observe my progress up the canyon. Often I would not return for many months, and yet it was always there waiting. I thought that perhaps there were many scrub jays in the area, and any one of them could be appearing on that branch. But as the years went by, I was able to detect the subtle shape and wing coloration of this exact bird. One day, after I had been absent for an entire year, that scrub jay greeted me in the canyon with its entire family. They all aligned on a branch and began singing the sweetest song. They were welcoming me back to their home. Eventually they came to my house in the desert. They entered with caution at first, but soon found comfort inside the large central room. Making an Ally of a wild creature like the scrub jay took a lengthy process of abstract communication, and the occasional real addition of a few tasty walnuts.

The arrival of your Ally will always come at a portentous moment on your walk. Pay attention to the timing of its arrival. What were you thinking about at that exact moment? Did it appear in the air, on the ground, or even on your person? Stop your internal debate and let the Voice of *Seeing* remind you that this is an affirmative gesture and nothing on your walk is a mere coincidence. You are learning to be patient. Each species lives in a particular time frame, slower or faster then the human assemblage point. By stopping your progress and becoming still, you are allowing their world time to dominate. The animal will recognize that you are acting out of the normal movement patterns of a human. If you get down lower, or simply freeze, they are interested. They recognize that you have stepped outside of the human Known.

Not-Doing: Seeing the Red Grid during the Day with Closed Eyes

Sit or stand in a location that offers a wide vista of blue sky. Close your eyes and look *directly* into the sun, so that the brightest area of the sunlight is *seen* through your closed eyelids. Take your right hand and with fingers slightly apart, begin *crisscrossing* your flattened palm across your closed eyes as close as possible to your face. Find the best hand position so that the open fingers *cast shadows* on your internal screen. Take your left hand and begin to wave it across your eyes in a cross rhythm with the right hand at the same time. Both hands are now casting distinct moving shadows over your closed eyelids. Concentrate on the visual field inside your closed eyelids. Alternate the fast crossing of the shadows so that you begin to *see* multiple small rectangles of light and dark inside your eyes.

Speed up the pace of your hand waving until you begin to *see* a demonstration of the *Red Grid:* the pattern of light and shadow grid lines and small red rectangular shapes moving under your closed eyelids. This not-doing is a practice run for the actual *seeing* of the Red Grid with open eyes in the sky. With the bright light of the sun as the source of illumination, the skin of your eyelids will give the patterns a red coloration. It is more then just a perceptual oddity. To the Toltec perceptual magicians, the *Red Grid* was a demonstration of the luminous body of light fibers that surround the globe of the Earth. Practicing this not-doing with closed eyes is a preparation for performing the exercise with open eyes.

Not-Doing: Seeing the Red Grid, Honeycomb and Sparkles during the Day with Open Eyes

Now open your eyes and look into the blue sky, but not directly at the sun. Begin the same open hand waving close to your open eyes. Do the motion with rapid intent. Now drop your hands and continue to *gaze* into the blue sky. You will *see* the Red Grid. It will be a "softer" version of the interior grid that you could *see* with the closed eye practice run. Continue to *gaze* at the sky until the reddish quality of the lines slowly becomes a shade of amber. The change in saturation, from red to an amber hue, was named by the ancient Toltec: the Honeycomb, the second stage of sky *gazing*. The lines of the luminous grid will become a soft golden hue patterning the blue sky with hexagonal prismatic shapes.

The Toltec believed that the honeybee was a carrier of the spirit. The method of their hive making and the delegation of work tasks made for a successful group of warriors. By closely observing the honeybees as Allies, the Toltec could *see* that through the use of their compound eyes, they could detect another world inside the flowers revealing the source for making honey. The bees became the Toltec symbol for enhanced *seeing*. The hive and the manifestation of honey made the bee a trusted Ally.

The third and final manifestation of the Red Grid was a not-doing named The Web of Sparkles. Through intended *seeing*, and in combination with magical movements, the grid structure dissolved into a pulsating burst of falling stars following a pattern set by the original grid lines. Each of the lines became a lightning pattern with a ball of intensity at each end, falling through the blue sky of the day. The perceptual magician was *seeing* the final manifestation of the luminous energy field around the globe. It was a moment of great joy, because the apprentice was *seeing*, in the purity of the light flashes, the actual workings at the electrical junctures or synapses of the luminous shell around the globe. The whole blue sky was a network of flashing lines. *Seeing* the Web of Sparkles was the culmination of *sky gazing*.

Not-Doing: Flattening Steep Ground

Go to a "steep" city sidewalk or a "steep" path in nature. Begin walking from the *bottom* of the steep grade up to the "top" high point of the sidewalk or trail. Keep your eyes focused just in front of your feet. You are concentrating on the area "right in front" of your walking feet as you progress up the hill. You can occasionally look up to validate that you are staying on the path, but immediately look back down. Your eyes always immediately return to the ground "right in front" of your feet. Hold your *gaze* in this position as you walk forward up the hill. *See* the ground in front of your steps as "flat". Your *seeing* will eliminate any mental information about the path being steep. In the process of keeping the eyes concentrated in front of the feet as you go up the path, a magical process occurs in your *seeing*. The normal bodily fatigue of climbing a steep grade is gone. This not-doing is tricking reason. The body is only sensing flat ground and the body feels only the minor exertion of a regular walk on level ground.

The Toltec used this not-doing to help the people of the tribe overcome the hardships of going long distances over steep mountainous terrain. This not-doing also relied on group trust. The lead warrior was responsible for putting the group on the proper trail by keeping their eyes open and scanning all the bends and turns of the path, and verbally instructing the following members, so that the remaining group behind them could use this not-doing technique, and keep their eyes looking downward.

Without trust there could be no unified group of perceptual magicians. Just as the honeybee hive functioned due to group alignment, the Toltec apprentices had to learn the necessity of sharing skills and energy. In order that trust was fostered, the teacher worked with more then one apprentice so each had the opportunity of *seeing* the specialties of the other. One apprentice found that their skill was locating a Larger Power Spot, while another was able to *see* the energy of a plant, or another the correct path through the mountains. Working as a group allowed each person to hone their personal skill and practice its effectiveness with the others in the group. Each apprentice learned to abdicate to the other's specialty when the immediate situation called for that particular skill. Decision-making was therefore deferred to each in their turn, and group trust was built in the process.

When I walked with my children we would designate one of them to be the guide for the day. It was the guide's responsibility to lead our group into unusual locations; to find something in the city or nature trails that would be a revelation to the rest of the group. One day, I was walking with my three children to the top of a mesa in Chaco Canyon, New Mexico. My oldest son was the guide. At one point he taunted his younger brother to lift up a large flat rock as a dare, because he said there was a rattlesnake under the rock. His younger brother met the challenge. He also trusted his brother and knew that he would not intentionally place him in danger. My older son had been his guide and was no doubt just testing his courage. He walked over and lifted the rock without hesitation. Instead of a rattlesnake there were two sticks of long incense under the rock. We were all stunned by the magical appearance. We all knew that my eldest son had not placed the incense under the rock.

The magic was in his not-doing choice of that particular rock, which he picked from the multitude of available rocks in this landscape. I often wonder, at that young age, if he was *seeing* energy from that particular rock. His adolescent dare with an impeccable sense of creativity, as our group guide, was really about the personal power of his *seeing,* which he had transformed into a brotherly challenge about a rattlesnake. He was *seeing* how energy flowed in the mountains that day, and although he would forget this event in his later years, for me, it was a moment grooved deeply into my Wheel of Time.

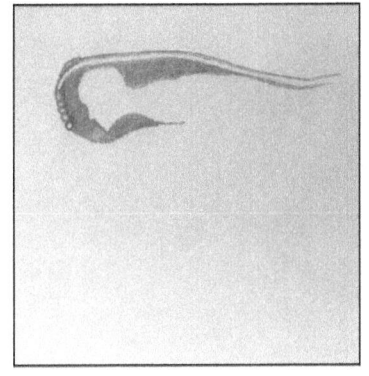

VII
The Directions: Personal Wings

Through their investigations over time, the Toltec could *see* that the physical location where a person was conceived was not only the beginning of their walk on Earth, but the entrance point of their "direction" for life. The place on Earth where conception originally occurred, determined the Latitude around the globe that was the individual's Largest Power Spot. Those imaginary Latitudinal Lines running East and West were a boundary framework where the person felt at home, and where power naturally flowed to them. A person had come from the Unknown through a portal in the energetic field of the parents who were making love. This doorway was not random, and although the person was picking the parents solely through the energy being generated, they were also at the right place at the right time.

The Earth would become their new home, and they had picked a particular passage in the luminous cocoon of the Earth that was directional and specific. From the small Personal Power Spot to the larger landscape Power Spot to the largest Power Spot within the entire circumference of the Globe, these were the micro to macro areas that would always emphasize the clearest awareness for an individual.

When an apprentice sat in a small circle of stones on their Personal Power Spot in the midst of wild nature, they were instructed to *see* that they were also part of a gigantic circle that extended around the entire planet. When their nature Personal Power Spot was located in the latitudinal circumference where they were conceived, the chances of direct answers to their intended *wish* would be magnified. Omens and agreements would occur with greater frequency in that particular band of possibilities.

When a female apprentice asked the benefactor to clarify her direction, she wasn't talking about her profession, but in the Toltec tradition, every person is not only part of the human mold, but each mold has a directional imprint. In the case of females the directions are classified as the Four Winds: the harsh wind of the North, the warm wind of the South, the light breeze of the East, and the pushing wind of the West. They were taken to their personal Power Spot. The teacher guide would *see* omens and agreements in the landscape that translated into one of the four cardinal directions: North, South, East or West. Once the apprentice had the knowledge of their Direction, a clear life path was revealed. All their future decisions took on a framework that easily contributed to a proper life. They were given a key for understanding their temperament and personal position in relationship to the larger Earth.

....

Your daughter is an adult. She is *wishing* to find her Direction. All these answers can be found in the rocks and plants of her Larger Power Spot. It is time to move into the realm of knowledge that is reserved for individuated adulthood.

You are walking with your daughter into her Larger Power Spot in the high mountains of New Mexico. The landscape is patterned with sections of red clay alternating with areas of white sand scattered with rough rocks and Pinyon trees. The radiating canyons starting in the high surrounding mountains cut from water runoff meet at a central lower mound. It rises up in the terrain like a small extinct volcanic shape. The Voice of *Seeing* says:

"There are constant decisions in life. After solving one problem, another is sure to arise. In the jungle of life, you are lost without a clear map. Knowing your birth direction stops internal debate, removing doubt and replacing it with confidence."

Your daughter is a grown woman. She has asked to know her Toltec Direction. She trusts you to *see* her clearly and point her in the right way for a successful life walk.

You climb up the steep sides and reach the rounded crest of the mound. It is a winter day and patches of snow remain under the shadows of the Pinyon trees. You instruct her to draw a circle in the soft sand with a stick, using the area that she feels proper. You watch her scan various locations until she is comfortable with her choice. She then places a stone in each of the four directions along the circumference of her lines. Her circle is located at the crest of the mound with a three hundred and sixty degree view of the entire red tinted mountainous landscape. You pay attention to the direction and quality of wind when a slight breeze begins to blow from the South. You go to a position where your daughter isn't distracted by your presence, but you can *see* her and all of her actions. She stands at the center of the circle and voices her *wish* to find her "direction". She says:

"I have come to this Power Spot to find myself. I am ready to learn my "direction", which will help me to know myself."

She steps over to the rock in the northern position of her directional wheel. She also faces outward toward the North. She waits in silence. You are intending to *see* any omens that validate a northern disposition. *Seeing* exterior omens is a direct presentation from the outside that disrupts the proceedings. This disruption could come with a sudden sound by an animal or a visual appearance of a leaf that floats into the circle, or any number of unusual occurrences that instantly punctuate and/or illuminate the beginning *wish*.

You are the *seeing* witness, while your daughter is the *seeing* receiver. She takes her time, waiting patiently to make sure that the North is empty of any agreements or omens. Nothing occurs energetically on the outside or within her own feelings so your daughter then steps over to the Eastern rock on the periphery of her circle. Again nothing occurs that would validate that direction. She experiences the same lack of an exterior omen, or interior feelings that would validate her western predilection. Finally your daughter steps over to the last possibility and faces the southern direction.

You are *seeing* beyond the young woman you have known since birth. Your own *seeing* must not be influenced by familiar prejudices. *Seeing* without looking. You close your eyes. Your daughter's frame appears as an after-image in your interior visual field. You can *see* a warm quality within her frame, as if the sun had created a halo around her. You feel a warm wind blowing from the South. You open your eyes. The setting sun has outlined her body with an amber glow. She is smiling, as if to acknowledge the sun's heat as a blessing in the coldness of this winter day. Your daughter yells:

"This is it, the South!"

You instruct her to leave the circle and walk toward the South, scanning for something that validates her feelings about this direction. As you walk with her into the South, she scans the sky and the Earth intending to *see* an omen. As she reaches the outer perimeter of the hill, she reaches down and retrieves something. When she opens her hand there is a small orange object fluttering against her palm. On closer inspection, it is a small remnant of a butterfly wing. She holds it up to the sun on the horizon, bringing the thin membrane into vibrant brightness. The Voice of *Seeing* says:

"Like a butterfly escaping its cocoon, a person who knows their direction is freed to fly without hesitation into the wider world. They know where they are going. And more importantly, why they are going."

You sit together *seeing* the amber sun create lines of bright light that traverse through any cloud barriers. The lines of the world spread out from the sun to connect all that exists under its influence. You inform her of all the characteristics of a Southern Woman from the Toltec perspective. The Southern Woman wants clarity and uses her words in a direct manner and expects the same from others. She is sometimes shy when speaking in front of a large group of people. But she is always warm and has love at the forefront of all her interactions.

Your daughter gets up and places the butterfly wing under the rock at the center of her stone circle. And then she says:

"I place this for safe keeping. This fallen wing reminds me how short our time is here. I must use my time wisely and fly free to meet that challenge."

As you walk down the mountain, your thoughts turn toward the future. Your daughter will use her warmth and direct intentions to illuminate her own children apprentices in the coming years. Her gait has an extra bounce of confidence. She is leading you now and runs down the path. When she stops to look back, you *see* her body momentarily vanish in the brightness of the last rays of the intense setting sun.

Not-Doing: Seeing Your Direction

Go to your natural Larger Power Spot. Position yourself at the center of an area where you have a clear view in all four cardinal directions. A raised area or a higher ridge that allows views in all directions will facilitate your success with this not-doing.

Draw a circle on the ground with an 8' diameter and place a rock in each of the four major compass directions at the perimeter of your circle: North, South, East and West. Stand and face outward at each direction, for the length of time it takes to feel a positive connection or a negative connection. Take your time to observe the landscape in front of you. *See* if anything stands out that makes you feel connected to that particular direction. If that first direction does not feel right, move on to the next compass position. Face outward at each direction in turn until you notice a positive feeling in one of the four directions, and/or until an omen of validation presents itself. Wait. Take the time to *gaze* into the landscape after you decide on one of the directions. Notice the harmony in the elemental composition of the landscape that spreads out before you. *Gaze* into the arrangement of color and light and shadow. You are collecting *seeing* information and aligning with the manifestations from the outside world.

When you have visually selected the direction that appeals to you, walk forward into the landscape in that chosen direction until you find a "validating omen": a special stone, a feather, or an animal Ally. As you walk in that direction you are continuing to sense the positive nature of your choice. The feeling of being connected to that direction should be intensifying as you walk further. Stop when you are completely sure of your "direction". Scan the landscape from your chosen position. You are *seeing* with your entire body; *see* the sounds from the birds, *see* the flora and fauna as contributing to the resolution of your *wish*, feel the direction of the wind. When you understand that you have validated your "direction", stand triumphant at this outer most boundary and say out loud a thank you to the spirit, to the elements, to all the animals and people that have contributed to your life walk.

Once you have discovered your personal "direction"… that direction never changes. It is part of the mold you were given at birth, and lasts till your time on this planet has ended. You are connected to mother Earth through your "direction", established when you first entered into this world. Your "direction" is a reflection of how you interact with others and the larger world.

Now that you know your personal "direction" you can explore its significance in the larger scheme of things. Discover where you were conceived (not necessarily the same place you were born), then trace the Latitude around the entire globe for a map of your Largest Power Spot. The latitude area of your place of conception will reflect your "direction" in relationship to the entire Globe. The cities within this global area are those that most clearly reflect your temperament. The climate and landscape between these Latitudinal Lines are the best-suited places for your success in life.

The Toltec could *see* that females of the human mold had four distinct patterns cut into their mold. These patterns were the genetic makeup of the four general types of female energy. Although each female had nuances of all the directions, one particular "direction" was always dominant. Finding the particular dominant "direction" placed the candidate within a band of knowledge that had been collected through practical validation over millennium.

The East is the place of the rising sun and fosters domestic responsibilities and family. The Eastern Woman is good at completing a designated task, but does not improvise well. The North is the place of the mind. The Northern Woman performs her tasks with ruthless intention and creativity, but her presentation is often cold. The West is the place of the setting sun and resolution. The Western Woman is mysteriously alluring but is also prone to deep empathy. The South is the place of body and sensual sensation. The Southern Woman takes command with a heartfelt style but can also disguise her cunning nature.

The knowledge of your "direction" forms a comforting structure. It defines many of the personality traits that one uses in confronting the rigors of the Known world. Much of the guessing on how to maneuver in the world was taken out of the equation with this directional knowledge. For the females in the Toltec Lineage, it was the wind of purification blowing away the questions about who they were, and replacing hesitation with the ease of intended action.

If the Toltec benefactor was guiding a male into the knowledge of their "direction", the techniques were similar, but the types of male activity and temperament were classified in a different syntax. The North was the place of the scholar. A Northern Man was the organizer of information, but prone to self-importance and risk taking. The East was the place of action. The Eastern Man was quick, but was also non-committal. The West was the place of ritual. The Western Man was dedicated to the mysterious, but sometimes indulgent. The South was the place of beauty. The Southern Man was a creative force, but needed restraint.

In both the male and female "directions", there is an overriding theme of being able to work with others who compliment your "direction". A leading Southern Woman will work best with an Eastern Woman and a Western Man as assistants. An Eastern Woman, when she is in a leadership position, will work best with a Northern Man and a Southern Woman as her assistants. The leading Western Woman works in tandem with the Southern Man and the Northern Woman. A leading Northern Woman is best assisted with a Western Man and a Southern Woman. In each of these arrangements, it is the leadership "direction" that dictates the other directional associates.

In order for group-trust to build, it is necessary to solve the riddle of the "directions" and begin to perform with precise acts of impeccability. The map of the "directions" was part of the ancient Toltec system of *seeing* how energy flows within the larger group for peak performance. Without this map, time was wasted with sloppy associations; the Toltec worked as if there was no time left before having to leave this beautiful planet.

The practical application of knowing a personal "direction" for my daughter came when she was intending a life partner. One day we sat together and reviewed her past boyfriends. We discovered that each of her three previous serious relationships had already developed her directional map. She had indeed dated a Southern Man, a Northern Man, and an Eastern Man. All of these lengthy friendships had given her a clear understanding of the types of men from a Toltec perspective. Each man had shared the best of himself from his singular "direction". The Northern Man had often challenged her fears by having her jump from airplanes. The Southern Man had taught her about joy and indulgence, and the Eastern Man had started her journey with an adolescent spark of real friendship. We both intended that the final puzzle piece would be a Western Man, the best compliment to her direction and the best suited for a long-term life partner for a Southern Woman.

The Toltec apprentice is given many gifts from the land. Their interpretation is subjective for every person, because nature holds unlimited creativity in all of its power gifts. Receiving the knowledge of your "direction" while standing on the landscape does not mean that it will be an easy transference of that knowledge, when back in the chaos of the world below the Power Spot. It is advisable to reflect on the Toltec saying: "You can take the apprentice to the water, but you can't make them walk on it." Giving the best of myself didn't mean that my daughter would accept all that her guide had to offer. As we walked down the Mountain that day, I reflected on the hardships of relationships and the search for love.

When we first walked together in a Toltec way in the urban park, so long ago when she was a young girl, our "wish" that day was to *see* "love". "Love" at that time represented for her the connections between school friends, her family and household pets. Now that my daughter was a grown woman, "love" meant finding a suitable partner. *Seeing* someone who will "love" you for a lifetime is not available to most people. Their choices for a suitable partner are based not on *seeing* but on "looking" at exterior appearance, a compatible personality, common interests, or final resolution formulated on some other fashion of the times criterion.

The ancient Toltec used *seeing* for determining partnerships that could stand the test of time. They would bring the couple into a Power Spot and observe the luminous cocoon around their physical bodies. If the *colored dust* around the two candidates was compatible and mixed together in a harmonious fashion, then their union got the approval from the perceptual magicians of the tribe. They trusted their *seeing* of energy as the final arbitrator for "love". They could *see* that this energetic connection would continuously foster awareness, which for the Toltec was the ultimate reason for two people coming together.

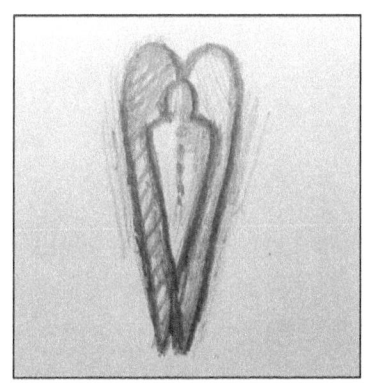

VIII
The Family: Last Wish

When the Toltec magician brought an apprentice to the ritual caves in the remote mountains, they brought them there to *see* the final puzzle piece of the perceptual mystery; to *see* the energetic makeup that connects everything in the universe. They were bringing them to the cave to be in complete darkness, the best place to view the *colored dust*. In the complete darkness of a deep cave, deep inside mother Earth, they would begin to have visions, formed from the sparkling dust that occurred in the absence of light.

The *colored dust* was the amalgamation of swirling defused pinpoints of color within the atmosphere of the deep void of the dark cave. The apprentice would hopefully form personal realizations from the chaos of shifting light patterns and begin to understand that *"looking"* was the result of light, and that the culmination of *"seeing"* was the result of the absence of light. And that "looking" was also the process of putting together information from the surface of objects, while *"seeing"* was the process of reading the light emanating from within everything.

The Toltec apprentice was left alone in the dark cave for as long as it took for them to understand the *colored dust*. Since this was the crust of the perceptual magician's path, the teacher was waiting outside with anxious anticipation of a successful outcome. It was the teacher's hope that the apprentice would understand that *seeing* was an abstract activity in its purest form, and that all the categories in the Known world were only various personal descriptions of the arrangement of the *colored dust*. It was the teacher's hope that the apprentice would *see* their own interpretation of the arrangement of lines and images as only a reflection of their own temperament and life experiences. It was the teacher's hope that eventually the *colored dust* would be *seen* without any personal filters, that it would be *seen* as pure energy. The apprentice would finally *see* that the emanations of the Eagle are woven through everything on Earth, activating awareness with the fibers of the Eagle's "will". These pinpoints of light had their origin with the Eagle's *wish* or bestowal of awareness. Through that bestowal, everyone would have the chance to *see* and become aware. And that was the ultimate intention of the spirit.

The moment the apprentice left the confines of the cave and entered into the bright sunlit landscape, or if they exited the cave in the night, they could *see* the pinpoints of light emanating from every tree and rock. They could *see* into the mystery of life and into their own bodily frame. They could *see* that their own hands were made of *colored dust*, and when moved across the blue sky of the day, or the dark night sky, a pattern of after images traced the progression of their movement through the air that was also the *colored dust*. Everything was a human description of something that could never be defined perfectly, because all the logical, reasonable words fell short when describing the Unknowable. The apprentice walked into the world of energetic *seeing*, and they were no longer apprentices. They had become *Seers*. It was the last and final gift of the Toltec teacher.

....

As a Toltec Father I had reached the end of my teachings. My children were adults now, and I had done my impeccable best to prepare them for life. The precipice edge of the Known path had been reached through the perceptual knowledge of the Unknown, and now there was only a vast panorama of infinite space stretching out into the Unknowable.

You have finally arrived back home with your three children. It has been a lifetime walk and they are all adults now. You remember all the walks together, over hard city streets and upon soft desert paths. You remember the first *seeing* adventures in Power Spots; when your daughter was able to *see* love in a river, and your sons were able to *see* power in the gifts of stone. Your wife greets everyone at the front door and ushers you into the original family Power Spot where your children were all born, the house that you built. As a Toltec Father, you have *seen* that love lasts forever when voiced as an intended *wish*. The spirit has also made a *wish*; a *"wish"* that resonates beyond the confines of the Known duality and resides eternal in the Wheel of Time. The spirit has *wished* for every family to *see* their connections beyond the troubles of the temporal past, and accept that "love as awareness" never ends. This is a walk into the eternal unity that remains as a lasting *wish* from the spirit. The Voice of *Seeing* says:

"You have listened carefully and followed the dictates of the spirit. Now there is one final wish. It is the long lasting wish that encapsulates the entire art of seeing. It is the wish of seeing everyone beyond the syntax of the Known world."

The Voice of *Seeing* is recommending for you to share with your family the end result of the mastery of *seeing*. It is something that unites them to each other and to all other people and places and things in the larger universe. The Voice of *Seeing* is suggesting that your family *see* together. The Voice of *Seeing* wants everyone to *see* how energy flows in the universe; to *see* each other as the same pulsating pinpoints of energy that make up the reality of the Unknowable.

Your children are all grown and moving into their own worlds; these moments together will become rare in the future. So you become the Toltec father guide for one last time and turn off the bright lights in the family room. You remember the great feeling of guiding your young children into *seeing* many different worlds throughout their lives, from the city of people to the Kingdom of the Birds. You recapitulate the moments of heartbreak and jubilation that occurred within the larger life walk on the curved path toward awareness. And you say out loud:

"The super power of *seeing* is available to every person but not everyone takes the chance to have a chance at *seeing*. All of you accepted the challenge I presented when you were very young. Long ago, I *wished* for all of you to be aware individuals. It was my intention to show you magical not-doings of perception, not only because it was creative play, but also in order for you to reach awareness. Each of you is my fulfilled *wish* that you become aware humans; humans who contribute to other people's awareness through their own *seeing*. Each of you is now a complete magician of perception."

Your children stand tall as warriors of perception who have battled their own frozen "looking" and joined the ranks of *Seers*. And then you take the millimeter of a chance, and ask if everyone is ready to fly on the wings of perception one last time before we traverse different life paths. Everyone nods in agreement.

You all stand together in a circle at the center of this family Power Spot: a circle of silhouettes in the dim light. You *see* that all your family is united in *seeing*. You *see* four strong silhouettes of the people you love become fluctuating frames holding the mystery of existence, the slow moving particles of infinity… the pulsating *colored dust*. The central body core of each of your family members is a nucleus of the mysterious fabric of the universe. You *see* your daughter, your sons, and your wife as magical beings made of billions of swirling pieces of starlight. You *see* the *colored dust* extend to everything in the dark room, rotating within everyone and everything. And then your daughter makes her final *wish:*

"I *wish*, that no matter where we all go in life, that at the end of our walks, we all *see* each other again in the next world after death."

And the room becomes a gigantic sea of *colored dust*. The walls and the floor fall away, becoming the same swirling particles of colored energy, the filaments that are the fabric of all things. And in this moment of *seeing*, our walks over all the landscapes of learning culminate in absolute stillness. A profound peace encompasses the atmosphere. You are floating in silent knowledge, linked together by the energy of intent.

Not-Doing: Seeing the Colored Dust

Go to a location that is dark. It can be a dark room, or it can be your Power Spot at night, or inside a deep cave; somewhere that has very limited light. The darkness is essential for extreme *seeing* without looking. *Gaze* into the darkness with relaxed vision. Now *still* your eyes in one section of the darkness.

Eventually the dark space before you will not be just a black area. If your *seeing* is developed and you meditate long enough with *stilled* vision, you will ultimately *see* that the darkness has a texture and a movement; it is made up of *colored dust*. The term *colored dust* is used by Toltec perceptual magicians to define the slow moving infinitesimal pulsating fabric of the universe that is made of all the subtle pinpoints of all colors, the subtle moving pattern that is all patterns, the lightness that is not bright light, the dark that is not completely black.

Another not-doing technique for helping to *see* the *colored dust* is to cup your hands around your eyes like binoculars; each hand is curved around each eye to limit the view to a small aperture. The cupped tubes around your eyes will focus your view in the darkness and eliminate any other visual distractions that might interfere; perhaps when your eyes adjust to the limited light you begin to perceive objects in the environment that distract from a clear unobstructed view. With the help of the hand binoculars you can now close one eye and view the atmosphere with the other open single eye. Then switch from eye to eye as you *gaze* into the *colored dust*. The "single eye" viewing helps prevent any "doing" fatigue and assists with clarifying the phenomenon. Some perceptual magicians focus on the pulsating web-like nature of the *colored dust*, defining this movement as a series of gyrating spider webs. As you begin to detect the *colored dust*, you will understand the difficulty in trying to define its appearance with words. It remains as the Unknowable.

As a very young child, I was able to *see* the *colored dust*. When I went to bed at night, I would raise my small hand and begin circling the dark atmosphere between my prone body and the ceiling of my room. At the time I assumed that everyone could *see* the water-like eddies that could be stirred with my hand waving. I had no context for what I was *seeing*, except the nomenclature of an eight-year-old boy. One of my favorite pastimes was playing with very small figurines or action figures. So, when one night I called my sister into my room to observe the swirling particles, I named them the "little men". With my rudimentary vocabulary, I described the small particles as tiny men running all around in the darkness of my room. My sister stood there for a while and then announced that she didn't *see* anything unusual.

With that first adventure in sharing *seeing* with another, I realized even at my young age, that a teacher couldn't predict the abilities of the student. At that moment, at the age of eight years old, I entered into the realm of a teacher of perceptual magic. I realized that "looking" was the accepted form of viewing the world, while deep "*seeing*" was a latent ability for most people that had to be activated by a teacher guide. My mission to share this deeper possibility with others began that night with my sister.

Not-Doing: Seeing the Colored Dust in a Person

Go to your place of power in the nighttime with another person or a group of people. It can be in the wild natural landscape or in the city park, but it must have limited distractions from outside light. Street lights in a public park, or even a bright moon lit landscape can be a limiting factor. If you *wish* to perform this not-doing in a room, turn off all the lights, pull the drapes and cover any openings that allow extraneous light to enter into the room. It is not necessary to have complete darkness, only the absence of most bright light sources or reflective surfaces. If you *wish* to perform this not-doing in nature, pick a location that is surrounded by shadowed bushes or a dark rock wall that can be used as a backdrop.

Have someone stand in front of you, approximately fifteen feet or more away, so that you can easily detect their whole body frame as they face you in the darkened room or natural landscape. *Gaze* into the center of their body and *still* your eyes in one location around the area of their chest or around the solar plexus. Become aware of their whole body outline; although there is limited light, your eyes will eventually become sensitive to the person's general silhouette or body frame.

Begin to *see* the *colored dust* circulating inside their silhouette. Remember that the *colored dust* is luminous but not bright, it is all colors but has no stationary hues. Your *seeing* can develop this nebulous atmosphere into understandable symmetrical patterns. Often the chaos of movement becomes a set of evenly aligned rows of interlocking circles. As humans, it is natural for us to use geometric shapes and lines to define the star patterns into zodiac signs, and to organize the Unknown qualities of our universe with mathematical equations and graphic geometry.

You will be *seeing* something that is not possible to define completely with reason because it is energy. The label *colored dust* is also flexible; it is only a short Toltec description of something that cannot be described with words. The Toltec also called it "the fabric of the universe" or "the glutinous atmosphere". They believed that they had detected the very substance that held together everything on the Earth, from the transparency of air to the density of stone. When they *gazed* into a rock wall it was no longer solid, but a mass of moving particles, and they could even penetrate the thin floating top layer with their hands. When they *gazed* into a shadow, they could *see* the *colored dust*. When they viewed a person, they could *see* that they were made of the *colored dust*. Everything in their world was connected with these fibers of rotating light.

After centuries of exploration concerning the *colored dust* by Toltec perceptual magicians, the mystery of the fabric of the universe was resolved. Teams of *Seers* found that the pinpoints of light were the ends of the emanations of the Eagle. Energetic bestowal through the endless lines of light, originated at the Eagle, and circled back, forming a continuous untangled loop. The bundled emanations went through everything. The perceptual magicians were simply *seeing* the bright ends of the thin tubes, where they naturally formed a screen or barrier available for their perception. A *Seer* was *seeing* the "frame" of luminosity around everything; the luminous shell was defined by the bright sectioning of the never-ending tube emanations of the Eagle.

One evening my oldest son invited me to visit a volcanic park in the hills of Northern California. We walked in the dark to a set of large boulders that ringed the rim, blown from the ancient caldera at the base of this highest point. He was interested in this area as his Personal Power Spot, and wanted me to verify his *seeing*. As we stood about twenty feet from one another, I heard the Voice of *Seeing*:

"The time has come to show your son the intricacies of the colored dust. He is a Seer now, and in this place of his personal power, a place of fire from the center of the Earth, a father will pass to his son the continuation of the lineage of the Toltec."

I instructed my son to *gaze* into my mid section from his distant position, and tell me what he was *seeing*. He easily elaborated on what he was experiencing. He spoke of *seeing* the *colored dust* for the first time in a cave he had discovered by the ocean. Now this energetic field was more active inside my human frame, and radiated upward to form a column of light into the night sky. More importantly, he understood that I had waited till he was thirty years old to open this final lesson in the art of *seeing*.

I wanted him to experience me as a father first; someone who was part of the Known world, participating in all the activities of a suburban man raising a traditional family. I had been waiting patiently to present the totality of myself to my son when he had become a man, encountering his own petty tyrants, his understanding of love, and his participation in the battlefield of Tonal life. Now he was an adult perceptual magician. At this moment, standing on the rim of an ancient caldera, a tear fell from my eye as I realized he had accepted his position as the next holder of the Toltec lineage of *seeing*.

THE FAMILY: LAST WISH

IX
The Next World: Past, Present, and Future

The Toltec believed that there was a method to maintain individual consciousness after the death of the physical body. They knew that old age and the eventual death of the physical body was a gate that no one could avoid, only perhaps postpone, and they accepted this Known reality. But simultaneously they took on the challenge of being able to maintain their Energy Body with its consciousness after the end of their physical body. They were able to *see* that the Energy Body was a vehicle that took dominance while they were dreaming. They knew that dreaming was a bridge to another world that could only be accessed with the non-organic Energy Body.

They deducted that the worlds accessed through dreaming were the logical extension of their consciousness without the need for a physical body. This Energy Body was also *seen* to exit the physical body of a dying person, and had to be going somewhere. The dominant feature of the Energy Body was its ability to *see*. The completed Energy Body didn't need arms or hands or legs, because it interacted in dream worlds through the intention of its eyes. With this knowledge, they began practicing with their Energy Body to collect information of dreamt worlds through this second set of eyes, long before death was eminent.

The Toltec used their Energy Bodies to go to remote locations in their dreaming; real worlds that required practical knowledge for survival. They used their "will" as fuel to project the Energy Body into distant landscapes that they named: the Sand World, the Water World, and the Air World. These new worlds could be visited in "group dreaming" for validation and corroboration as to their makeup, and suitability for habitation after death. They began to construct edifices of intent in remote locations of the universe; places that would be their new homes after they permanently left the Earth World. In this way, they had escaped the fate of a man who was left without any tools for survival at the time of leaving this Earth at death. The Toltec became the masters of their own fate.

The magicians of antiquity accomplished the feat of building a gigantic dome in the Belt of Orion. They named this place of refuge, the Dome of the Nagual. It was created through the "will" of many Energy Bodies, and manifested through the intent of their eyes. The mastery of *seeing* with the Energy Body was the greatest accomplishment of those magicians. Not only had they collected valuable information and manifested art with their *seeing* on Earth, but they had also used their dreaming eyes to construct a practical structure in a distant star field. The Dome of the Nagual was built with their *seeing* intent. It was the place to meet with others who were also on their definitive journey.

In the process of exercising their Energy Bodies, they calculated that "time", as a method for judging success, was flexible and inconsistent. The time it would take to walk to a distant mountain with their physical body was significantly longer than the time it took to reach the same location with their Energy Body. To test their discovery concerning the flexibility of time, they would have an apprentice sit with eyes closed at the topmost point of a mesa that dropped off into a deep canyon, and have them practice jumping into that abyss with their Energy Body.

The teacher would place selected objects or arrange stones in a pattern that the apprentice could only *see* if their Energy Body arrived at the bottom of the canyon. The time it took to *see* the arrangement was instantaneous. The apprentice could be in two places at exactly the same instant; sitting quietly at the top of the mesa, and also at the bottom of the canyon. The use of the Energy Body to travel to remote locations in an instant was tested with further and further destinations, until the stars themselves could be reached immediately.

The Toltec had *seen* that the power that rules all life on Earth was the immense power that they named the Eagle. This power was too removed from their system of classification so they visualized it as a gigantic eagle. The Eagle was the guardian of energy, and not to be confused with God or the spirit. By using their Energy Bodies they had traveled to the end of the Known universe to glimpse the Eagle standing guard at the last gate of dreaming. They could *see* that in order to pass by the Eagle and go on their definitive journey to the other side, they would have to leave their mold of man within its gigantic talons. They would return the gift of life energy received from the Four Nagual Fathers… the "three seeds" of their awareness. The Eagle would accept this borrowed gift and allow them entrance with full consciousness into the next world, on the other side of its all-*seeing* eyes.

....

You are dying. You have returned to your Personal Power Spot on the land for one last time in this life. The time has come when all your Earth doings and not-doings are complete. You have been called by the Eagle to return the energy it lent to you at birth. You lay within a perfect circle of placed stones gazing upward into your next destination. In the silence of no internal dialogue, you make your last *wish*. It is the *wish* to ignite all the fibers of your own *colored dust* with the emanations at large, the lines of energy that connect directly with the Eagle. Each fiber is a memory of your time on this Earth. Each fiber ends in the burning light of collected knowledge. High above you in the clear blue sky are the witnesses of circling condors. Two deer come to graze next to your body. They all know you are leaving. The trees cast long shadows. The Earth remembers your footfalls. And the wind gives you one last moment of life. You feel the burning of your light fibers being pulled in all directions. The ends of your *colored dust* fuse in bright radiance with the lifelines of the Eagle. You burn from the fire within. You take the last step, the step out of your physical body and into your Energy Body... permanently.

You are riding to the top of a building in a bubble of fluffy light. Overhead is a huge dome that extends from horizon to horizon in all directions. You have arrived at the Dome of the Nagual. The bubble elevator lets you out at the roof garden. It is arranged like a landscape you have visited in the past, with a lawn and big boulders. Only the colors are different. The lawn has an amber hue and the rocks are bluish green. Angelic beings are sitting on the lawn writing notes and placing them inside holes in the textured cracks of the large rocks. You are curious as to the nature of this activity and you go over to one of the boulders. Inside a small crevice is a rolled message that you remove. When you unroll the note, you are shocked when you read the inscription.

It is something that you yourself wrote when you took your three children long ago to the top of a skyscraper back on Earth. You are moved to tears when suddenly all of your life work condenses into this one moment. You instantly *see* the past when your children climbed that boulder to place a branch at the top, as you wrote your *wish* and rolled up the paper to place in the stone. Time has coalesced into one dream; the past and the present have blended together into this future. You fall down into the amber grass that cushions your descent into this specialized groove in the Wheel of Time. The note reads:

"All my wishes have been granted."

You are an accomplished perceptual magician. You have traveled far and practiced often unfurling your wings of perception. There is no frozen assemblage point on your luminous cocoon. This is the note you wrote, and it is still held within your heart. The Wheel of Time acknowledges your intention for infinity. As a perceptual magician you will carry its message to the ends of the universe. You have peered deeply into the makeup of the Earth and arrived with your Energy Body to this next world.

The Voice of *Seeing* has been a trusted Ally for your whole Earth life. The Voice of *Seeing*... once heard... remains with you after death. Its counsel continues after you have left this earthly body. As it has helped with important decisions in your waking life, as it has directed your actions in dreaming, so it will continue to advise your Energy Body within the realms of the Unknown. It will remain with you for another million years. You are going to explore new worlds now, worlds that are beyond the gate of death. The rest of the universe is available to you. And then you hear the Voice of *Seeing*:

"Within the Wheel of Time there is an option to survive after physical death with complete awareness. To be granted this option, a person must leave a tribute. This tribute is a duplicate of their collected perceptual knowledge while alive on the Earth. You have left your duplicate in the form of this book."

And suddenly the dome above you becomes a sheet of pulsating *colored dust*. Out to the horizon of infinite space, as far as you can *see*, the entire universe is made of *colored dust*. And you look down at your own frame, to *see* it is also made of the same pinpoints of modulating light. Everything is *seen* as blending slightly into everything else, and a subtle wind is creating currents of ripples that unite everything in this gigantic pulsating sea of lights. You glide forward with your Energy Body. And then you *see* the Eagle forming within the shifting patterns of *colored dust*. At the base of the Eagle are the Energy Bodies of your three children who have come to visit you in dreaming from Earth. And your daughter peers into your eyes with unbending intent and she asks:

"What is your *wish*, father?"

And you answer with the only sentence that completes the cycle between life and death; the only answer a perceptual magician can give to another perceptual magician when all their *wishes* have already been granted. And you say:

"Lets go further."

And then, as the atmosphere becomes a shimmering conglomeration of unending possibilities, you and your children float effortlessly over a rainbow bridge, past the gigantic presence of the guardian Eagle… out into the next *seeing* adventure that fulfills the ultimate *wish* of the spirit.

Not-Doing: Seeing Everything

Lay in your Personal Power Spot. You have traveled far on your life path and have come this day for a final *wish*. Now you lay down on your back, relaxed, with your eyes closed inside the field of the second attention, the place of silent knowledge.

You have performed your not-doings as a perceptual magician with impeccability. You have shared your *seeing* with your children and with those who you have loved on this long life walk. Now it is time to project your Energy Body out into the far horizon. You are leaving the confines of this planet, and traveling to your next destination with your Energy Body. Your physical body stays where you lay on the Earth as a subtle anchor, ready to pull you back on the thinnest of fibers when you are through visiting this outside world. Your physical body is the anchor mooring you back to Earth; it connects with a single fiber to the Energy Body that enables a gentle return back to your tonal life.

This is not the time to permanently leave, only a not-doing practice run. Because every person must leave this beautiful Earth eventually, the Toltec believed that practicing for their definitive journey was mandatory. Without some familiarity with the process and some experience with the destination, a perceptual magician would face unnecessary trauma and confusion after death. Your Energy Body destination this time would be the Dome of the Nagual.

Everyone has a conception and visualization of Heaven. Every person and every culture has their specialized filters concerning images of Heaven, in order to make sense of the indefinable. Our sacred books are a humble human attempt to translate the Unknowable into the Known. These filters are a comforting collection revealed over time to perceptual magicians around the globe, and then written in sacred texts that foretell a future life after death. The Toltec gave their apprentices a way to verify their teachings concerning death, by giving them the real experience of the destination and the practical tools of navigation. They brought them to the Dome of the Nagual before physical death using their Energy Bodies.

When you arrive at the Dome, you will be greeted with an angelic guide. This being is an inorganic Ally manifested in an angelic form. Your internal dialogue doesn't exist in this place. Only the Voice of *Seeing* remains in all the inhabitants.

There is no linear time involved with travel for the Energy Body. You will arrive at the Dome instantly. When you touch down, observe with your closed eyes the scenes developing around you: the floating angelic beings with large eyes, the beautiful cloud formations, and the architecture intended by the Toltec of antiquity.

Perhaps your last *wish* is to arrive at Heaven. Perhaps your vision of Heaven is connected to your favorite vacation spot: a sunny beach with swaying palm trees and celestial music. Feel the warmth and safety of the environment. Perhaps your imaginings of Heaven involve past companions who are there to greet you in a splendid garden. Although your filters are unique to your experience and culture, the unifying theme for all travelers will be the feeling of arriving home at long last.

See the other occupants and their contentment within this world they had *wished* for, when they were still alive on Earth. *See* if you know anyone, perhaps a loved one who has died long ago. Remember that you have used your Energy Body to arrive in this place, and there is no time. You are simultaneously in the past, present and the future. Some of the arrivals may have left their Earth bodies long ago; some may arrive in their dreaming from the present time. *See* all those whom you have loved arrive in this Heaven, some of them prepared to go even further on their definitive journey into the larger infinity.

In the Toltec Map, the Nagual leader is responsible for assembling a group of warriors for their definitive journey after death. When they meet with their Energy Bodies in the Dome of the Nagual after death, they have fulfilled the final not-doing of their particular lineage. They had arranged before death to meet on the other side, prepared for even further explorations of worlds upon worlds. They have *wished* to *see* other worlds and explore as far as possible into the wonders of infinity. In order for these perceptual magicians to survive in the vast predatory universe that will be filled with new obstacles and challenges, they have solved the riddle of their Core of warriors. The riddle of the Core included: group trust, understanding each person's special skills, loss of self-importance, and complete and ultimate "love as awareness" for each other.

When I am walking in nature and come across a bird feather I am reminded of the transitory nature of life. I *see* this feather as a power gift. Just as the bird losses a feather, or a snake sheds its skin, so we also will eventually return our own physical bodies back to the Earth. I usually hold the bird feather up to the bright sun. What I *see* through the many barbs of the vane wing is a rainbow. The light of the sun creates iridescent effects across the barbs of the feather. I *see* all the colors of the rainbow. It never fails to bring the power of happiness to my heart. It is a gift with an interior rainbow. It is the most incredible appendage that allows for flight. And it usually comes across my path at just the perfect moment… at a time when I am thinking about my own mortality.

X
The Recapitulation

The ancient Toltec had an expression that removed doubt from the apprentice… it was: "We believe because we want to believe". This was their way of stating that the universe was a mystery, and they accepted their *seeing* as the gift that would set them free even though, that in it self was inexplicable. The perceptual magician was making a *wish* with that statement.

Young children don't innocently believe in magic, they believe because they are complete and they can already *see*. In the common nomenclature, there is fascinating labeling for children who can *see*. When I was a child, *gazing* out the window during a particularly boring lecture, I was told by the teacher to stop "day-dreaming". The phrase "day-dreaming" has magic at its core, but has been twisted to denote a negative connotation. Another that is often distorted into a negative meaning is the phrase "it's just your imagination". Again we have a general dumbing down of the original magical intent of this word, which was an avenue toward *seeing*.

The Toltec were given a treasure map that they wanted to believe, and it enabled them to disperse to every continent on the Earth in search of the treasure called awareness. They had different names in Egypt or in Japan, but they all were recognized as the holders of *seeing*. Toltec is the general name for the artists of MesoAmerica, but the lineage began on the island of Atlantis. With the sinking of their homeland, the Atlanteans traveled to new continents. They sculpted monuments, and designed great temples; they observed the natural world and brought back gifts that helped to create civilizations.

I am a Toltec Nagual. My Toltec name given to me by grandfather don Arturo is Ayakel. Ayakel is a transmitter of dreams. A Toltec Nagual is the lineage's specialist in the art of *seeing*. It took me many years to discover my bloodline, and to be found by a benefactor who validated my Nagual position, named me, and guided me into the Quetzalcoatl Lineage of perceptual magicians. One of the traits of a Nagual is their unbending intent to add to the existing perceptual knowledge of the lineage. I accepted my responsibility to continue my heritage and to bring this information to you in the form of writing and illustrations.

I have been your Voice of *Seeing* during the reading of this book. In order to present this material, I have become that clear voice inside your head. This is my historical role as a teacher of the methods of the Toltec. My last gift in this book will be the following recapitulation of my own journey to arrive at this transmission. The recapitulation is the Toltec term for reviewing the energetic scenes from ones life, through *seeing*, that directly pertains to the acquisition of awareness. They use closed eyes or open eyes to visualize and stimulate the memories of past events, and they take back any lost energy through breathing, or through intensely reviewing the scenes with their arts. In this book I have recapitulated my knowledge with writing and illustrations; using these arts as a propellant into the art of *seeing*.

It is the process of openly reviewing my life with my children that gives this recapitulation a bridge to your own life as a parent and as a guide. My children were my guides, and the validators whether my *seeing* was in fact effective for others. Until a teacher has an apprentice in the Toltec tradition, they have no clear mirror for their own *seeing*. Each apprentice reflects a facet of the teacher's own personality and serves as a beacon of recognition through their example. My children gave me the freedom to guide them, and I in turn, gave them the freedom to be themselves, and that is an ultimate power gift of love.

The core of the lens of your eyes is the only body part that maintains its original cells for your entire life. Once the fibers of transparent proteins circulate around the central core and establish the lens of the eye, you awaken as a perceiver, using this same lens for life. All the information that you collect with your eyes is filtered through this transparent membrane. You begin life by gathering information from the outside world, moving it through the lenses and then into your internal memory.

When you go beyond "looking" at the world, the lens of perception bypasses reason, and saturates the brain with energetic *seeing*. *Seeing* is the art of using the Known world as a platform for moving into the Unknown world of perceptual magic, and eventually arriving at the gate of the Unknowable. You are in one world, but through *seeing* you become acutely aware of the division between accepted reasonable thinking and disguised ultra-sensing phenomena. You are a natural perceiver, but *seeing*, points out the areas where our senses have taken on a frozen role due to societal constraints. The Known world is constantly reinforcing acceptable perceptual information so that we can relate to our environment in an agreed upon format with other humans. The art of *seeing* involves expressing your latent perceptual abilities with creative not-doings.

One of the first not-doings I performed with my apprentices was having them close their eyes while I guided them walking around a familiar city. I was their trusted guide, making sure that they lifted their feet to avoid a curb on the street, and generally watching out for their bodies. I guided them to areas or objects in the environment and instructed them to quickly open their eyes and then shut them again. With this process they were *seeing* only a fleeting image. The city that they had been accustomed to viewing took on a whole new dimension. These momentary glimpses made the Known into the Unknown. Most apprentices were stunned by the small details that had been excluded from the usual sights in the city. They all validated the ability of these rapid snapshots to activate internal creativity once their eyes were closed again. The process of opening their eyes for a quick glance extracted the mysterious from the mundane.

The perceptual magician silences their mind, and intends to *see* further, to *hear* more intricacies, and to face a more mysterious and complex world. The intended magician moves into a wider perceptual framework in order to *see* the way the first humans saw the world, without contemporary "glosses". They purify and cleanse the senses of those layers of taught behavior. They *see* with all their senses, not just the eyes. One of the first magical tasks practiced by my young children involved altering their sense of hearing. We would stand together in nature and cup the palms of our hands behind our ears to form a type of sound magnifying wall. This cupped palm enlarged the range of our hearing. We heard nature's world of sounds magnified. All the sounds around us were enhanced. We were hearing like a deer, with its larger ears. When we put our hands back down we could *see* how much of the available sound world vanished from our normal experience.

The techniques of *seeing* are used in both the urban city and the natural countryside. The flying Condor of the natural desert, or the deer in the deep forest are the equivalent of the helpful cab driver in the city. Both environments and their characters are the scripts and actors in the Play of unlimited creativity. The only requirement is the continual use of your imagination. The effectiveness of the art of *seeing* is measured by the unpredictable events that we have called "magic", occurring throughout the walk. Without the internal validation through the Voice of *Seeing,* the Play would not be entirely successful. It is the Voice of *Seeing* that punctuates your *wish* with a path "marker"; a brief solidified coherent description of the illusive magical moment. This internal description of our *seeing* is the tangible art that we bring back from our walk.

The success of the Play is measured many years later when your own child is an adult. If these not-doings do their intended job, then your children will carry the tools of creative perception into whatever profession they enter as an adult. I have no doubt that these not-doings are effective as a catalyst for creative thinking or better yet, for creative *seeing*. I have tested each of the exercises with my own three children, and the children of friends, and have seen the results of our work together when they have reached adulthood. But be forewarned, your life will never be the same. You and your children will join the esteemed ranks of perceptual magicians throughout time; those who have dreamt and manifested great art, those who have designed great cities, those who have invented new products that help humankind.

My oldest child is the character template for the young girl in the stories. She is now the art therapist for the most prestigious high school in the area. She helps the best and brightest to become not only fully functional adults, but also the next generation of creative individuals. My oldest son is a filmmaker and a stone mason. He has joined the lightest of human arts, filmmaking, with the heaviest of human arts, the sculpting of stone. My youngest son is a dog whisperer. He can communicate with canines so that they become "service dogs" for those in need of a furry trusted Ally.

If you asked any one of them about their childhood, they will pay tribute to the days of play together with a *seeing* script. My daughter might share the moment we found a dried leaf in the forest that had a small insect hole at its center. We placed a drop of water from the nearby stream over the center hole. Then we looked through the hole, through the suspended water drip at the forest ground. The water membrane acted as a magnifying glass. We saw a new world through that small water covered hole. It was a natural lens for *seeing* in a new creative manner.

All my children are artists of perception. They are helping the community to *see* themselves and to *see* their environment as something that still holds the mysterious. They are the next generation of perceptual teachers. My daughter uses her *seeing* skills to guide high school students into becoming functioning adults. My oldest son uses *seeing* to create magnificent works of art. And my youngest son uses *seeing* to communicate with animals. All my children practice the art of *seeing*; they walk within their contemporary world as formed *Seers*. Their game board is now the entire world.

When I walk in the landscape these days, I remember the times when I placed a ring of stones, or left a carved branch leaning against a tree for someone to find, and use as either a walking stick or a magic wand. I hear stories about people coming upon a circle of stones while hiking and being activated by the magic of the discovery. Sometimes I return to those markers that I have constructed, only to find that a rock is now out of place, either blown by the elements, or moved by a running deer. I replace the stone every time. Sometimes the site has been enhanced by strangers who leave another ring of stones inside the one that I had designed, increasing the complexity of the design. Sometimes the site is so old that moss and sediment have almost buried the work into obscurity.

Finding the remnants of another human's experience, left in a wild place or in a crevice of a city wall is an exhilarating experience. These unknown magicians of perception have gone beyond the Known and worked diligently to decorate our environments with tokens of their experience in that location. They are in the lineage of cave painters. The ancient sites of indigenous cave art still intrigue our modern sensibilities. We wonder why these artists of antiquity left the Known consensus of their tribe to enter into a special place in order to leave a painting of their adventures on a cave wall. The impact of their *seeing* is still felt today. Their intention to humbly depict the Unknown is the best of our humanness.

The concrete results of *Seeing* is when we are leaving art for other people to discover through their own *seeing*. There is nothing more exciting then coming across something created by another human who spontaneously used the natural elements to make a piece of art in the woods or in the city. We marvel at the effort they took to stop their walk in order to create something beautiful for others to find.

I remember coming across a cave in the high mountains of northern California. It was located in an isolated area where very few other humans ventured. It had become my Power Spot. Inside the small cave was a group of wooden twigs bundled together with braded tree bark. Each of the lengths was identical to the other, and the bark had been carefully removed to reveal the smooth interior of each twig. The sculpture must have been made some time ago, as it was half buried, and the wind and rain had created a mossy patina over the entire grouping. I brought the bundle out into the sunlight for closer inspection. There were five separate twigs. It made me think about simple offerings that activate the larger imagination. I thought it was very beautiful, and wondered if it represented something about "family" to the artist. I imagined that each one of the twigs represented someone in my own family.

As I stood that day in my northern California Power Spot, within the textured shadows of the bay trees, I suddenly remembered every thing about that bundle. Long ago I had been the magician who tied those twigs together and left the bundle in that cave.

The bundle of twigs represented myself, my wife, and my three young children, tied together for infinity. It was a moment in the Wheel of Time that I had forgotten until that instant. I was moved to tears. I placed the bundle back inside the cave; but not before I made an intention. I said in a loud voice:

"I *wish* that I forget once again!"

When you are moved into remembering a *seeing* act that you performed long ago, you are experiencing a tangible recapitulation in what the Toltec called… a groove in the Wheel of Time. This is the same exultation that led primitive man toward the exploration of new possibilities in human awareness. Memory was linked to *seeing*, and the collection of information became the obsession of the evolving human race. They became the artist, the scientist, the doctor and the magician of time and space.

The first lineage of perceptual magicians developed many of the skills that have now become independent specialties, but were once bundled together under the title of shaman. The original magician was a maximum *seer* of perceptual oddities. They taught their children how to do the perceptual shifts that in later eons would be labeled as "creativity".

Seeing is the return to the original creativity of "*Seeing* how energy flows in the universe.*"* You are given permission to walk out of the confines of the

Known and into the field of a Power Spot. In this place, the mystery returns and you realize that you also are a mystery. You give yourself over to the challenge of attempting to solve the big riddles of life: why am I here, what are my gifts to share, where am I going. Your senses come alive again. You intend to revive the clearest vision of yourself, your children, and your world with the pure tools of perception. The world is alive again. You have become the original creativity of your own child.

When you guide someone else into the landscape of the Unknown, you have given a gift that will last a lifetime. The duality of plans or organized fun can often deplete the original motivation behind getting together. One of the favorite holidays for children is Christmas. It is the time when presents are piled beneath the Christmas tree awaiting the fateful day when the wrappings are torn asunder and the present inside the box is at last revealed. Watching my own children unwrap a present in great anticipation, only to have their hopes of something else in the box dashed away with that first view, was an agonizing moment. In order to prevent such disappointment we used the art of *seeing* and a not-doing.

Each of my children had a special day with their father before Christmas. On that day, they could do anything they *wished* and buy any present or presents they discovered. My children set their own limits and devised their own gaming field. One of them would pick a particular street, or perhaps we would begin in nature to later arrive at a department store of their choosing. Everything during that day was a not-doing. We simply placed ourselves within the gaming field and let spontaneous creativity guide us into the mystery. Once the gifts were purchased, they were wrapped carefully and placed under the tree for Christmas day. When each package was opened, there was only great joy at finding the exact thing they had *wished* for inside the box. The clothes my daughter had picked were the exact size and design. My sons were never disappointed with their boxes, which held the perfect game or action figures. They had each utilized *seeing* to discover the proper environment with the intention of finding a magical gift that was the complete answer to their Christmas *wish*.

The other important part of this not-doing was the rearranging of the Known. We had taken a frozen holiday and turned it into a flexible Play with a perfect script of our own creative intention. The *doings* of Christmas involve anxiety and fear that a person will not like the present you have selected. We often run around like crazy trying to find the perfect gift for all of our acquaintances. Perhaps we have some inkling of their needs, but more often then not we come away from the store with thoughts of partial failure. In the worst case the person will even fake happiness when they see what was in the box, and everyone senses that the whole event was a charade. A not-doing reverses the negative into the positive by taking the stress out of the event and replacing it with intended *seeing*. There was only joy and success under our Christmas tree.

One day my son, who was six or seven at the time, decided that he was going to trap a wild bird. He set up a cardboard box tilted on a stick that was connected to a long string. His idea was to pull the string from his hiding position behind a bush when the bird went inside the box to eat the seeds he had left. He waited all day. He was determined to capture a wild bird, and his system had failed. No bird came to the box. He had constructed a magnificent *doing* from the place of reason, and yet success was out of his grasp.

I don't think that my son ever recovered from that event. It set him on the course of a warrior who realizes that the world consists of uncontrollable forces. He came to the perceptual magician's conclusion that it wasn't about winning and capturing a bird, and it wasn't about losing the chance to have a bird... it was only about the challenge. If you only "look" at the world, the duality of comparisons will dominate your thinking, right or wrong, win or lose, but a perceptual magician *sees* everything as a challenge; and accepting that challenge helps to form them into perceptual warriors.

The art of *seeing* is for those who *wish* for a moment away from the Known and moving toward the Unknown. It is an act that takes commitment to move momentarily away from the accepted modes of "looking". Perceptual magicians accept the challenge.

They get aboard the spaceship seeking new lands to explore. They arrive at a foreign shore and begin to *see* its treasures. They are *wish* makers, arriving with the intention of *seeing* the totality of their world as well as the totality of themselves.

From the moment I was born I have been blessed with the gift of *seeing*. The gift of sight is the most magnificent power gift in my life. When our sight fades it is a reminder that all super powers diminish with age. I have used this gift in every waking moment, and used my second sight in dreaming. It has truly been a trusted Ally till the end.

I have been a magician my entire life. I started as a magician of slight-of-hand parlor tricks and then moved on to esoteric practices of the Toltec perceptual magician. All along the way, I have used my *seeing* to unravel the mysteries of this Earth. In many ways this book completes my magical journey that first began when I *crossed* my eyes and saw two worlds as a young child. Throughout my adulthood I explored in paint and wood, in music, and performance, everything that I was *seeing* in this world. I became the primal shaman, combining artistic and scientific exploration. I became the alchemist of old, studying the makeup of the universe with the intention of making base matter into the gold of the totality of my *seeing*.

Like the alchemist, I have often failed to enlighten the general public with my offerings. However, the challenge of attempting the impossible has kept me walking down this curved path. I finished writing this book in a foreign land, far from my birthplace, far from the deserts and forests and the city streets that inspired these stories. At this moment I am far away from my children, my co-conspirators in these stories.

My travels had led me to a museum in Amsterdam, viewing a self-portrait painting by the artist Vincent Van Gogh. He became my Ally in that moment, and I heard Vincent say:

"My paintings are a gift of my *seeing* adventures on Earth. Each of my brush marks is a super power. My images are omens for those seeking answers to the difficulties of life. I gave my own life over to the challenge of *seeing*."

Vincent encouraged me to not lose my desire to succeed at revealing the gift of *seeing* with this book. Vincent was a Voice of *Seeing* whispering in my ear to not give up, because it is a warriors' challenge to present something new and beautiful to this world. Even when my internal dialogue attempted to slow my creativity with thoughts of defeat, at that moment, the Voice of *Seeing* came to my rescue with a very simple admonition:

"You have already succeeded. Someone, somewhere, at this very moment is thinking about your words. They are hearing your voice and its counsel."

One day, long ago in the Wheel of Time, I was walking with my young daughter through a deep ravine in Canyon de Chelly, Arizona. We had just crossed a small river and were sitting on a sand island with water on both sides. Downstream a Navajo woman in traditional dress was herding a group of goats. My daughter and I were opening a Play restaurant that served various mud pies. The sun was just beginning to set, so the entire canyon was bathed in amber light. It was a perfect moment. I knew that this magical image would stay with me for my entire life. Each person in this Play was perfectly cast, and the script was written by powers far greater than any human.

As the sun burst into its final bright illumination of the scene, I intended to hold this vision of Heaven forever. It was at that moment, when the world stopped, and my own internal dialogue as a father and a guide was silenced, that my own Voice of *Seeing* first appeared. And the Voice of *Seeing* said:

"You will remember this moment. All the images and colors will return whenever you review your life. I am here to help you understand, that you have heard my voice from the beginning. Often you thought it was your own voice. But it is a translation into your dialect. The source is outside of you. Some call it the voice of God; others call it the voice of the spirit. At this moment you are finally hearing the difference between your own internal debate and the clear voice from a higher source."

One day, in the far future when your daughter is a grown woman and you have long ago died to this Earth, she will be walking in wild nature. Her Larger Power Spot is a remote beach in Northern California. The tall cliffs with twisted pine trees form the eastern boundary, while the western direction is the vast blue crashing sea. She is walking south on the beach following her personal color wheel, picking up pieces of green seaweed, white sea shells, and collecting the petals of yellow flowers that grow wild in the almond colored dunes. When she arrives at her Personal Power Spot she arranges the power gifts on the sand into a mandala of radiating colors, with an inner circle of bright yellow daisies, and the white spokes of shells ending at heart shaped leaves, ringed in shiny seaweed.

She sits in silence *gazing* into the rolling waves. She takes out her *seeing catcher*; a small jaw harp that produces a vibration when plucked with the hand and activated with the breath. The sound of the *seeing catcher* moves her to close her eyes and extend her Energy Body toward the light of the setting sun. She flies toward the bright amber light through the red lines of the grid, using her "will" as fuel. And then she hears The Voice of *Seeing*:

"The intent of not-doing is to fly with the unimpeded wish of the spirit. The results of not-

doing is a natural ability to improvise spontaneously and create beauty. When the avenues of perception have been widened, then the apprentice is no longer an apprentice. They have blossomed into a teacher for the next generation of perceptual magicians. Passing the wand of creativity completes the unending wish from the Voice of Seeing."

The drone of her *seeing catcher* blends with the sounds of the waves, punctuated with the calls from the sea birds. Your daughter opens her eyes and *sees* a figure approaching from the north. It is still far enough away to only be a small shadow following the uneven patterns at the meeting edge of water and sand. As the silhouette gets closer, your daughter *sees* in this person an omen that fulfills her *wish* made at the beginning of this walk on the beach. He is a young boy. He is her oldest son. Your daughter has taken this time alone while her husband has been watching all three children further up the beach. When her son reaches the mandala and sits next to his mother, and puts his head in her lap, he fulfills the *wish* made this day by his mother, and that long ago was the same *wish* made by her own father, and the same *wish* that has lasted since the dawn of the Wheel of Time. In his young eyes is the fulfillment of the everlasting *wish* of every person who seeks the awareness of love when looking for another person. Her son says out loud:

"I wished to find you."

And your daughter understands that the past, the present, and the future unfold around a *wish*. She *sees* that the vast path of existence is fulfilled in the eyes of another. When the bonds of blindness have been removed, then the eyes of wonder are freed. And in that instant her own Voice of *Seeing* is the sound of the waves and calls of the birds, and the never ending world of light and shadow that told her:

"The secret to existence is being able to listen. "

When the words of humankind are directed toward sharing, and the eyes of humankind are focused on *seeing*, then the curtain that has concealed the nature of God will be opened to the unending Play of creativity. And creativity will write the script for infinity.

THE RECAPITULATION

THE END

www.ingramcontent.com/pod-product-compliance
Lightning Source LLC
Chambersburg PA
CBHW030612220526
45463CB00004B/1269